Storms Over Ireland

Aerial Actions 1913–1945

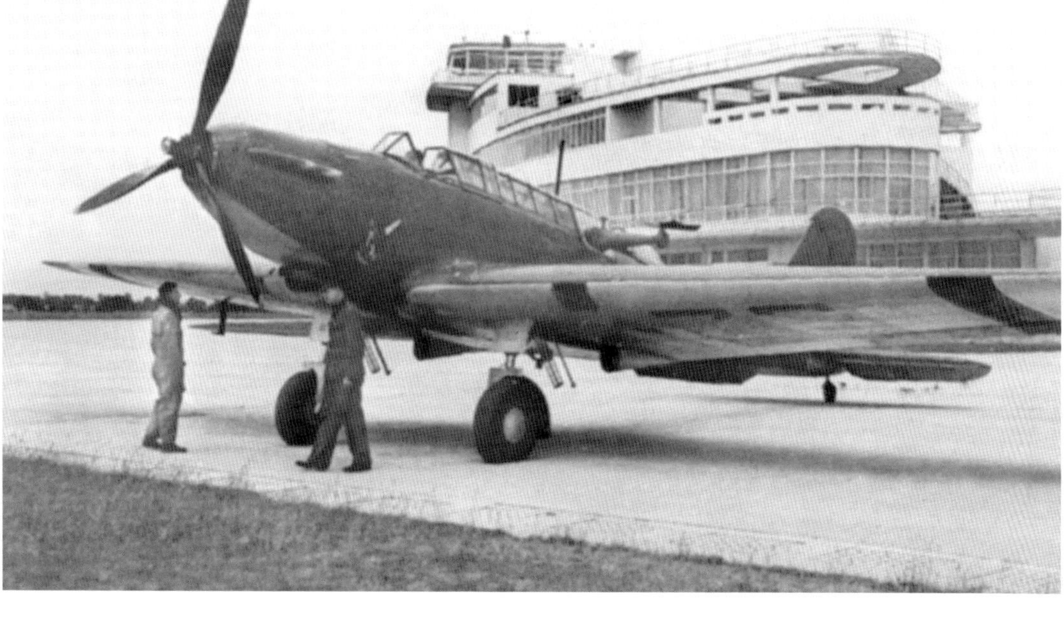

PETE LONDON

KEY Books

Front cover image: Hurricane I P5178 of 79 Squadron Royal Air Force (RAF) force-landed at Kilmuckridge, Ballyvadden, Co Wexford on 29 September 1940. Repaired by the Air Corps, it joined up as 93.

Title page image: In April 1941, the RAF's Fairey Battle TT.I V1222 landed intact near Waterford, and was subsequently pressed into Air Corps use as 92. At rest on Collinstown's apron (now Dublin Airport) during 1945, its forward fuselage bears artwork of Donald Duck holding a target, with bullet holes spelling out 'Get quacking'.

Contents page image: Air Corps Short Service Class Wings Day, Baldonnel, 1939. Back (left to right): 2nd Lts T J Russell, S Kelleher, A J Thornton, D V Cousins, M T Cregg, T P O'Mahony, H J Lee. Front (left to right): 2nd Lts M Maloney and M J Ryan, Lts P Swan and M P Quinlan, Capt T L Kennedy, Lt K T Curran, 2nd Lts T P McKeown and O W Lynam. An Avro 636 provides the backdrop. In January 1942, Lt Alan Thornton stole a Walrus amphibian and headed for France, intending to join the Luftwaffe.

Acknowledgements

This book has been many, many years in the making and would have been impossible to complete without the help and generosity of the following people and institutions:

Chris Ashworth, Jack Bruce, Sean Bryan, Barry Cole, Peter Cooksley, Tom Goskar, Padraig Crilly, Anne Grant, Rebecca Gray, Jim Halley, Rod Hall-Jones, Paul Higgins, Tony Kearns, G Stuart Leslie, Donal MacCarron, Eric Morgan, Ces Mowthorpe, Brían Murphy, Donnie Nelson, Aisling O'Riordan, Colin Owers, Pete Rosewall, Barry Saxton, Bernard Taylor, Lt Col Louis Treacy, British Library, British Newspaper Library, Documents on Irish Foreign Policy Project, Fleet Air Arm Museum, Imperial War Museum, Irish Defence Forces Military Archives, Irish Newspaper Archives, National Library of Ireland, National Museum of Ireland, RAF Museum, The National Archives (Britain), The National Archives (Ireland).

Dedication

This book is for my wonderful lady, Rebecca. She may not get round to reading quite all of it, but she's heard a lot about it – thank 'ee maid!

Published by Key Books
An imprint of Key Publishing Ltd
PO Box 100
Stamford
Lincs PE9 1XQ

www.keypublishing.com

The right of Pete London to be identified as the author of this book has been asserted in accordance with the Copyright, Designs and Patents Act 1988 Sections 77 and 78.

Copyright © Pete London, 2024

All images Pete London collection unless otherwise indicated.

ISBN 978 1 80282 862 7

All rights reserved. Reproduction in whole or in part in any form whatsoever or by any means is strictly prohibited without the prior permission of the Publisher.

Typeset by SJmagic DESIGN SERVICES, India.

Contents

Introduction 4

Part 1   British Rule, 1913–1918 5

Part 2   War of Independence, 1919–1921 29

Part 3   Civil War, 1922–1923 43

Part 4   The Emergency, 1939–1945 61

Appendix 1   Ex-RAF identities of Air Service aircraft, Irish Civil War 94

Appendix 2   Ex-RAF identities of Air Corps aircraft, Emergency period corresponding with the World War Two in Europe 95

Introduction

This book explores military aviation campaigns and actions in Ireland between 1913 and 1945. Until 1922 its scope is the entire island, which up to then was part of the United Kingdom. Focus then shifts to the Irish Free State's civil war of 1922–1923, and neutral Ireland's Emergency period coinciding with World War Two.

In 1913, the British Royal Flying Corps (RFC) made its first journey overseas, Ireland the destination. During the Great War, a substantial aerial presence arrived from Britain; military training airfields appeared, along with anti-U-boat forces comprising aircraft and airships. Germany's submarines were also hunted by Irish-based American flying-boats.

Directly after the war, the history-steeped question of Ireland's independence boiled over. In December 1918's general election, nationalist Sinn Féin candidates won a huge majority across most of the island. Rejecting Westminster's parliament, in January 1919 Sinn Féin formed the first Dáil Éireann (Assembly of Ireland), an entirely separate administration.

At once, the Irish War of Independence flared. As Britain sought to recover its authority, RAF aircraft helped search for nationalist guerrilla forces and safeguard Crown communications. Finally, following a truce, in December 1921 the signing of the Anglo-Irish Treaty heralded a new dominion: the Irish Free State. Hostilities ended, the British prepared to leave.

But the emerging country experienced severe internal conflict over the conditions set out in the treaty. By June 1922, the Free State's provisional government had entered a bitter civil war against an opposition that saw the terms of the agreement, well short of full Irish independence, as a betrayal of its cause. Nevertheless, with the treaty's ratification in December 1922, the new nation was born amid the fighting.

During the civil war, a Free State military air service took shape, engaging anti-treaty forces until the government's victory in May 1923. The following year, the Air Corps was formed from its predecessor. In 1937, the Free State's constitution was changed and the country renamed Ireland (Éire).

As World War Two erupted, the Dáil declared a state of national emergency. The Air Corps, attempting to support Ireland's foreign policy of neutrality, had scant resources to challenge warring nations' aerial incursions or oppose possible attack. Behind the scenes, efforts were made to obtain sorely-needed modern aircraft from Britain while seeking to preserve a particularly delicate diplomatic position.

Part 1
British Rule, 1913–1918

First Callers

Monday 1 September 1913, the coastal town of Newcastle, County Down: from the early afternoon sky came a raucous, unfamiliar noise. Staring over the sea, local people were amazed to see an aeroplane approaching. It set down on the sandy beach and rolled gently to a stop, the first military aircraft to alight in Ireland.

As part of British army manoeuvres based around Rathbane military camp, Limerick, that autumn, the RFC's 2 Squadron, based at Montrose in Scotland, sent six aeroplanes across the Irish Sea. Five were Royal Aircraft Factory BE.2a machines: Capt John Becke piloted serial number 217, Capt Charles Longcroft 218, Irish-born Lt Leonard Dawes 225, Capt Archibald MacLean 272 and Lt Francis Waldron 273. Capt George Dawes, born in Dublin and Leonard's brother, flew Maurice Farman Longhorn 207.

This was the RFC's earliest overseas deployment, and the aeroplanes took individual routes to Limerick. Only the Farman alighted at Newcastle; Dundalk's Royal Field Artillery barracks, Co Louth, and Curragh Camp army base, Co Kildare, were among other stopovers. At Rathbane, an RFC ground party had installed temporary hangars, tents, a workshop, spares and fuel. Four machines arrived

Left to right, Capt Longcroft, Capt Dawes, Capt Becke, Lt Waldron, Capt MacLean, Capt Tucker and Lt Dawes of 2 Squadron Royal Flying Corps (RFC). In September 1913, all flew to Ireland except Tucker. Behind the men is a BE.2a.

George Dawes and Farman Longhorn 207. For its over-water crossing, the machine received temporary underwing flotation bags. A spare wheel is attached to the struttery. BE.2a 273 sits behind. (Donnie Nelson)

on 1 September, Longcroft on 3 September, after suffering a damaged propeller shaft, and MacLean two days later with 272.

Participating in the manoeuvres were opposing 'Brown' and 'White' armies involving more than 20,000 troops, their movements resolutely observed by the aircraft despite varied weather including wind and rain. Valuable intelligence was provided by pilots returned from scouting while messages in weighted bags were sometimes thrown to the military below, there being no wirelesses fitted.

On 2 September, George Dawes and 207 left Limerick for Birr, King's County (later Co Offaly), seeking a landing field near the army barracks there. While setting down in gusts at nearby Sharavogue he broke an interplane strut, but a mechanic from Rathbane arrived by motorcycle and made repairs. Four days later, three BE.2as gave a display at Birr before rapt townsfolk.

Then, on 8 September, travelling from Rathbane to the Curragh, 218 and Longcroft crashed at Brosna, King's County. The *Irish Times* reported: '[218] landed in a hole, smashing some of the struts. It continued to run along, and got into a ditch, where further damage was done.' Longcroft wasn't injured, but 218 shattered its undercarriage and propeller. Three days later, Becke and 217 came down with a faulty petrol feed near Knockshegowna, Co Tipperary; the local blacksmith fixed the problem. Becke (217), Longcroft (with 218 repaired) and Waldron (273) visited Birr on 13 September, their machines guarded by local policemen.

The aeroplanes began their return home on 24 September. Leonard Dawes (225), Longcroft (218) and MacLean (272) made Montrose without mishap. Becke turned up the following day, 217 having crash-landed at Moyagher, Co Meath, where it was re-engined. Meanwhile, the Farman suffered fuel flow trouble at Roscrea, Co Tipperary, and further difficulties at the Curragh; it seems 207 never arrived back in Scotland.

BE.2a 218 travelled to Ireland with Charles Longcroft. The machine survived until the following year, crashing on 2 May 1914.

Charles Longcroft (seated) and Francis Waldron with a BE.2a at Rathbane.

Lt Rutter Martyn, who'd taken over 273, experienced engine problems near Ballyhornan, Co Down, and crashed. Martyn recovered, while damaged 273 was shipped home by sea. Accidents and breakages notwithstanding – typical for the time – the aerial force had underscored its military usefulness, despite makeshift operating conditions and difficult weather.

RFC and RAF

Amid the Great War's unprecedented conflict, in November 1915 the RFC returned to Ireland when No.1 Training Centre formed at Curragh Camp, a trade school for fitters and riggers. Dublin's Easter Rising of April 1916, the Irish republican movement's insurrection against the British government, didn't affect the training, though its aftermath contributed to an increase in support for independence.

Early summer 1917 saw much larger-scale plans emerge, for Irish flying training airfields, and a surveying visit was duly made by the RFC's (T) Maj William Sholto Douglas (later Marshal of the RAF 1st Baron Douglas of Kirtleside). Flying a BE.2c, Douglas identified three suitable sites near the capital at Baldonnel, Collinstown and Cookstown, Co Dublin, and another at Gormanston, Co Meath. Meanwhile, Curragh Camp prepared to host a day-bomber training station. Its first flying unit, 19 Training Squadron, appeared in December, with Airco DH.4s, Airco DH.9s and Avro 504s housed in canvas-covered Bessonneau hangars.

That month too, the nomadic and popular Air Services Exhibition arrived at Dublin's Earlsfort Garage. The attraction was organised by air-minded British socialite, the Countess of Drogheda. Exhibits included Sopwith Camel B5155, and as propaganda a captured Albatros C type, pieces of a wrecked Gotha bomber and remains from destroyed zeppelins. Aimed partly at stimulating recruitment to the flying services, proceeds helped fund hospitals treating injured airmen.

By spring 1918, work was underway on the new airfields. Each became a Training Depot Station (TDS): No.22 (Gormanston), No.23 (Baldonnel), No.24 (Collinstown) and No.25 (Cookstown). The TDSs were

January 1916: budding RFC ground crew at Curragh Camp's No.1 Training Centre.

Dublin, December 1917: the Countess of Drogheda's Air Services Exhibition. Among the exhibits were models of an airship and a parachute, Camel B5155, and a captured Albatros C type.

configured similarly, with paired Belfast hangars, a large repair shed and accommodation for all officers and men. At each base, 72 aircraft were envisaged – 36 Avro 504s for flying training and 36 DH.9s for bombing tuition. A target of 90 pupils would be present at any one time.

Actual aircraft strengths varied with deliveries and accidents, while DH.4s also appeared. Other types arrived in small numbers: BE.2s, RAF SE.5as, Sopwith Pups and Camels. Gormanston formed on 1 August, receiving A Flt 130 Squadron and C Flt 137 Squadron; Baldonnel on 1 September (B Flt 135 Squadron and A Flt 164 Squadron); Collinstown on 15 August (A Flt 156 Squadron and A Flt 163 Squadron); and Cookstown – by then renamed Tallaght – on 1 September (B Flt 136 Squadron and B Flt 164 Squadron). By October, all stations were operational under the 3rd Training Wing headquartered at Collinstown.

Additional modest landing fields were created around the country at Ballylin House near Ferbane, King's County; Birr; Castlebar, Co Mayo; Fermoy, Co Cork; and Oranmore, Co Galway. In October, the Irish Flying Instructors School formed at the Curragh with Avro 504Ks.

✠ ✠ ✠

Meanwhile, by May 1918, an airfield had also been established at Strathroy, near Omagh, Co Tyrone, hosting 105 Squadron on RAF RE.8s. Supposedly an anti-submarine unit, 105 undertook inland reconnaissances supporting Britain's security forces, observing nationalist rallies and areas of potential civil unrest. Serious public protest and further momentum for self-rule, driven by deep hostility towards the military conscription plan then being considered for Ireland, and by the arrest of numerous Sinn Féin members.

Prominent Irish nationalist William Gibson, 2nd Baron Ashbourne, at Phoenix Park, Dublin, on 4 January 1918, ready to view the city from aloft, his joyride organised through the Countess of Drogheda. Ashbourne's pilot was Capt Inglefield, the aircraft a DH.4 named *Firefly II*, which accompanied the Air Services Exhibition.

That May too, Fermoy received 106 Squadron (RE.8s), which also operated in a policing role. By August, 105's A and B Flts had moved to Oranmore, and C Flt to Castlebar. Detachments flew from Athlone (though the ground was found prone to waterlogging), Baldonnel, Birr, Collinstown, the Curragh and Fermoy. On 13 November, returning from Oranmore to Omagh, two 105 machines collided over Lissaneden, Co Tyrone; three of the four crewmen died.

Ireland's distance from Europe would have ensured no direct involvement with warfare, had it not been for Germany's U-boats operating in the surrounding seas. One of numerous countermeasures to that grave threat, though a minor one except for those concerned, in September 1918 Tallaght received 530 (Special Duty) Flt – ex-C Flt, 244 Squadron – equipped with the Airco DH.6 two-seater. These low-powered machines, their bomb-load negligible, flew paired coastal patrols over the Irish Sea and escorted the daily Dublin–Holyhead mail boats as far as the Kish lighthouse, seven miles off the capital. From time to time too, the DH.6s flew from Dublin's large public open space, Phoenix Park.

On the civilian side of life, Ireland's war effort included supply of considerable quantities of flax for manufacture of aeroplane linen, while numerous companies produced aircraft parts; Thomas Thompson & Son in Carlow made Bristol F.2B wings and struts. Belfast shipbuilder Harland & Wolff built complete machines under subcontract: 300 DH.6s (C5451–C5750), 300 Avro 504Js (C5751–C6050) and 300 504Ks (E301–E600). These were dispatched in crates. The company was also contracted to manufacture 20 Handley Page V/1500 four-engined bombers, five of which were test-flown and delivered just post-war from an airfield opened in May 1918 at Aldergrove, Co Antrim, as No.16 Aircraft Acceptance Park.

March 1918: Baldonnel's construction underway, footings for the hangars alongside temporary huts. Early in its life the station was sometimes referred to as Clondalkin aerodrome.

Cookstown/Tallaght takes shape, July 1918; part of the land is still being farmed.

Strathroy airfield, 1918. Visible are two Bessonneau hangars and between them a visiting DH.6, along with tented accommodation.

RE.8 C2378 of 105 Squadron at rest at Strathroy, late 1918.

Fermoy's initial layout, September 1918. The Bessonneau hangars were soon replaced by metal types, the tents by huts.

Avro 504K E403 *Mavourneen*, Irish for 'my darling', at Tallaght, 1918.

Top: An unusual visitor to Bentra's airship sub-station was FE.2b A5509, possibly over from Scotland. A naval type sits in the cockpit.

Above: Members of 105 Squadron and mascot line up with an RE.8, Strathroy, late 1918.

Left: Harland & Wolff-constructed Avro 504s under licence, including 504J C6001.

Fermoy, 1918: a ground crewman swings the propeller of a visiting TDS Avro 504J.

Right: The first Harland & Wolff-built Handley Page V/1500, E4307, at Aldergrove's No.16 Aircraft Acceptance Park. Appearing just before the Armistice, it didn't fly until 20 December 1918.

Below: Castlebar, late 1918: 105 Squadron personnel with one of the unit's RE.8s.

Left: RAF recruitment advertisement in Dublin-published *Irish Life* society magazine, 27 September 1918. Alas, the image used was of a German Albatros, probably derived from a photograph of a captured example wearing RAF markings. By that time, several such machines had been brought to Britain for testing or exhibition.

Below: Tallaght's Women's Royal Air Force (WRAF) contingent, along with a station nurse and canine chums, late 1918. The airfield site later hosted the Urney Chocolates factory.

Airships and Balloons

A much more substantial component of Ireland's aerial war against the U-boat menace, in 1917 Royal Naval Air Service airships began to arrive. These craft were of great endurance, allowing long patrols, and could carry substantial bomb-loads. To receive them, sub-station bases (also known as mooring-out stations) were created at Larne, Co Antrim (reporting to Royal Naval Air Station [RNAS] Luce Bay, a larger establishment across the Irish Sea at West Freugh); at Johnstown Castle, Co Wexford (reporting to RNAS Pembroke); and at Malahide, Co Dublin (reporting to RNAS Anglesey).

The bases enabled airships from the British 'mainland' to make their patrols, alight in Ireland, if required, to refuel and take on more rations, then continue their beats; in bad weather they provided safe havens. Larne, named after the nearby lough, was also known as Whitehead (the nearest town) and Bentra (the open countryside hosting the base). Similarly, Johnstown Castle was sometimes called Wexford, being situated around four miles south-west of the town.

Larne acquired wooden accommodation huts, and a portable canvas-covered shed in which to house one non-rigid Submarine Scout (SS: sometimes Sea Scout) airship type. The base's origins date to autumn 1915, but it was probably spring 1917 before Larne saw its first craft. Johnstown Castle went operational in May 1918, Malahide in July.

In spring 1918, construction began of a sub-station at Ballyliffin, Co Donegal. However, building was still incomplete by October and it seems few flights were made before peace intervened. At Ballyquirke, Killeagh, Co Cork, during September 1918, work started on a larger station with two steel sheds intended to shelter two Coastal Class non-rigid and two R-33 Class rigid airships. Despite the war's end, building continued until August 1919, but Ballyquirke remained unfinished and no flying took place.

A further scheme had appeared in 1916 for a rigid airship base on the eastern side of Lough Neagh, Co Antrim, some 20 miles west of Belfast. It was planned the craft would patrol over the eastern Atlantic and

Utilising a BE.2c fuselage, sandbagged SS.20 in Larne's canvas shed. The impression of an eight-bladed propeller is caused by the photographer having used a slow camera exposure. (Ces Mowthorpe)

if required, even the North Sea, but the station was never built. Proposals were also made for non-rigid sites at Castlebar, and Killarney, Co Kerry, as well as a station near Cork city for both rigid and non-rigid types; none were implemented.

┿ ┿ ┿

Larne's first airship arrived on 5 June 1917, SS.20 seeking shelter from poor weather. That month SS.23 also alighted with engine trouble, but made good on site. Late summer saw the base's shed borrowed by Mr Long, the farmer who owned the land, to store his hay. November's gales damaged the shed's doors; army repairers arrived, but spent their visit scoffing naval rations and left with the doors still unserviceable. Eventually a sail-maker put matters right. Suffering engine problems, Submarine Scout Zero (sometimes Sea Scout Zero) non-rigid airship SSZ.12 appeared on 3 March 1918, and again on 9 April with a damaged elevator. Besides SS and SSZ craft, numerous Coastal, Coastal Star and North Sea Class airships visited Larne. Personnel grew to around ten officers and 140 men.

At Malahide, the ratings lived in bell tents and the officers at Malahide Castle, a similar arrangement applying at Johnstown Castle. Neither base received an airship shed but at both sites trees were felled, eating into woodland to create sheltered areas where the craft could be safely moored. A scheme was drawn up for a permanent shed at Johnstown Castle to house three SSZs and a single Submarine (or Sea) Scout Twin type, but the Armistice halted this.

Johnstown Castle's first airships, SSZ.56 and SSZ.52, called in May 1918. In June, a hydrogen-producing plant and a bomb store were built, and in July a wireless station. Malahide and Johnstown Castle hosted various SS and SSZ craft, which continued their long, lonely patrols until the Armistice. Some of Ireland's airships stayed for lengthy periods, only returning to the larger stations for major overhauls or engine replacements.

Submarine encounters – even suspected – were very rare. In late May 1918, American destroyer USS *Wilkes* depth-charged a submarine spotted near the Saltese Islands, off Kilmore Quay, Co Wexford.

Fitted with a BE.2c fuselage, SS.23 at Larne with its ground-handling crew. By May 1918, unserviceable, it was scrapped. (Ces Mowthorpe)

Johnstown Castle: an SSZ airship secured in a clearing. The airships' lightness and considerable windage made them susceptible to gusts, the trees lending natural shelter.

Right: Malahide: an SSZ Class airship and attendants. In the background is a mooring clearing and tented accommodation. (Ces Mowthorpe)

Below: SSZ.52 with its ground crew at Johnstown Castle, 1918. (Ces Mowthorpe)

SSZ.58 is manoeuvred through a clearing at Malahide, 1918. (J M Bruce/G S Leslie collection, via Stuart Leslie)

A handling team walks a British kite balloon from Rathmullan down a nearby lane.

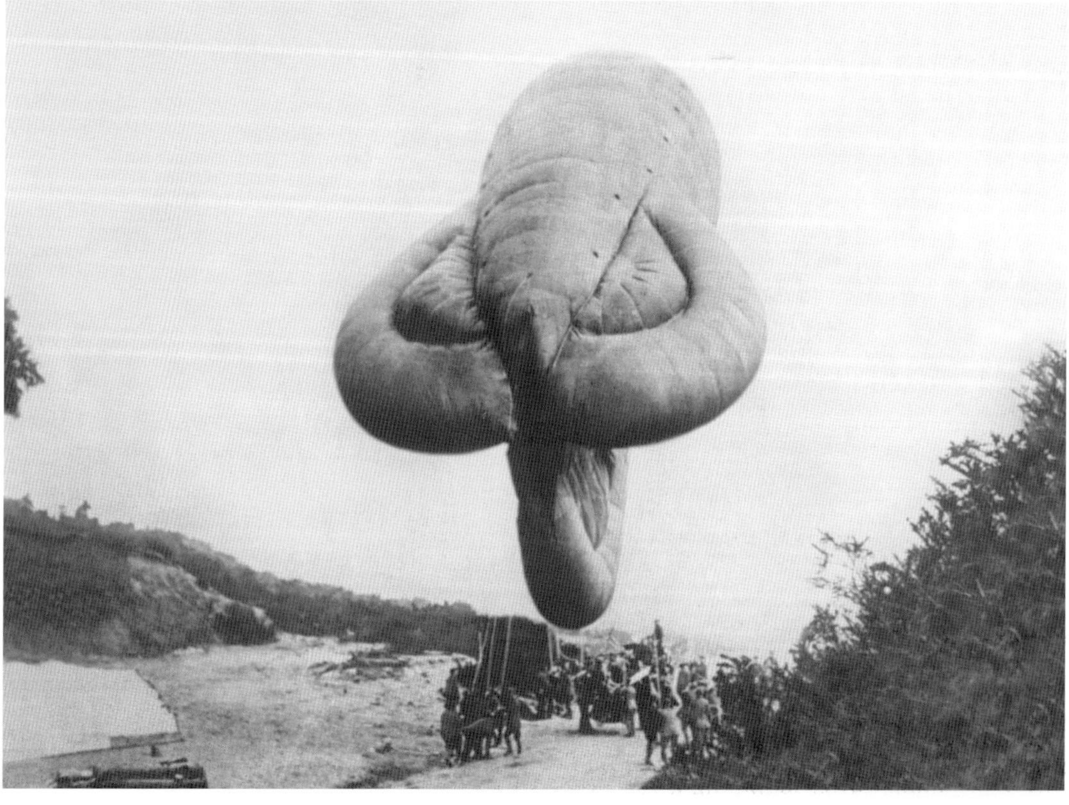

Berehaven's 17 Balloon Base hosted six balloon sheds together with accommodation huts. Oil storage tanks and a small jetty are also visible.

Johnstown Castle's SSZ.56 joined the fray, releasing bombs. However, German naval records don't indicate a loss. During May or June, near Tuskar lighthouse off Co Wexford, SSZ.56 was fired on by three surfaced U-boats using their deck guns. Presenting such a large target, it was forced to retire. Occasionally the airships came across itinerant sea mines, detonating them with machine-gun fire. But despite so few engagements, their deterrent effect alone helped inhibit the U-boats' effectiveness.

⊹ ⊹ ⊹

Tethered to naval surface vessels, manned kite balloons acted as lookouts to alert escorted Allied merchant convoys of nearby U-boats or lurking sea mines. The balloons were generally stored and maintained ashore. Ireland received two Admiralty-built kite balloon stations, both near centres of transatlantic convoy assembly. During the war, the country experienced a huge naval and merchant shipping presence; the balloons were kept busy.

Kite Balloon Station Rathmullan, on Lough Swilly's eastern bank near Buncrana, Co Donegal, went operational early in 1918, working with destroyers and sloops. Four canvas sheds each accommodated one balloon, and on its 14 acres Rathmullan housed typically 14 officers and 125 men. From 1 April 1918, the unit became known as 13 Balloon Base.

The 67-acre 17 Balloon Base was built at Berehaven, north of Bere Island, Co Cork. Six sheds, quarters for 20 officers and 150 men, and a hydrogen-producing plant appeared. In April 1917, the United States had entered the war and a year later the Admiralty transferred the base to the US Navy, which renamed it US Naval Air Station (USNAS) Berehaven. Indeed, 1918 saw America's Irish presence grow considerably.

US Naval Air Service

Seeking to increase Ireland's aerial anti-submarine patrols, early in 1917 the Admiralty had begun searching for sites at which to build flying-boat stations. Four locations were selected: Aghada, near Queenstown (later Cobh), Co Cork; Ferrybank, Co Wexford; Lough Foyle, just north of Muff, Co Donegal on the lough's western shore; and Whiddy Island, at the eastern end of Bantry Bay, Co Cork. Ferrybank, across the harbour from Wexford, became known by the town's name.

Following America's entry into the war though, it was agreed these bases would be operated by the United States Naval Air Service. The US Navy took over their construction – Irish civilian labour was also used – and in March 1918 the first materiel consignment arrived from America. Queenstown

Spring 1918: Queenstown a-building, two hangars nearing completion with footings for a third. On the water, a Curtiss H-16.

An American team manhandles a Curtiss H-16 at Queenstown. The machine rests on a detachable beaching chassis, and lacks its rudder.

became the principal and largest station, combining the functions of maritime patrol, aircraft assembly and repair, training, and headquarters for the bases' overall commanding officer, Cmdr Frank McCrary.

Each station received large hangars (Lough Foyle and Queenstown three, Wexford and Whiddy Island two), a waterside concrete pan, a slipway to launch and recover the flying-boats, and hutted accommodation. Queenstown was commissioned in February 1918, Wexford in May, Whiddy Island and gusty Lough Foyle – mostly surrounded by high hills – in July. Aghada Hall housed Queenstown's officers while at Wexford, nearby Ely House and Bann Aboo mansion houses were requisitioned.

The patrol area for Queenstown ran from Cape Clear to the western English Channel, while Lough Foyle covered the waters off the north coast, Wexford the sector east of Queenstown and the southern Irish Sea,

H-16 Bureau Number (BuNo) A-1076 on Queenstown's slipway with its sizeable handling crew. The aircraft later served at Wexford.

Queenstown: H-16 A-1076. Although a Curtiss design, this example was built by America's Naval Aircraft Factory. (Colin Owers)

and Whiddy Island the south-west approaches. The aircraft type flown from all stations was the substantial twin-engined Curtiss H-16 flying-boat carrying four crewmen. It was planned Queenstown and Whiddy Island would each receive 24 machines, with 18 apiece for Lough Foyle and Wexford. As well as anti-submarine work, the H-16s undertook convoy escorts, searches for missing vessels and training flights.

Lough Foyle began operations on 3 September, when H-16 Bureau Number (BuNo) A-1059, also allotted station identity LF-1, made a four-hour patrol. Its wings subsequently found defective, A-1059 received as replacements those of A-779/LF-5. Whiddy Island commenced work on 25 September, using A-1072 and A-1078. Five days later, A-1075 flew Queenstown's first mission; the following month that machine and five more made a total of 40 patrols. However, between 4 and 11 October three 'boats were lost after alighting at sea, two (possibly all) having experienced problems with their Liberty engines.

During mid-September, Wexford had received A-1074, A-1076, A-1079 and A-1087, followed by A3478 (to replace A-1074, lost after another emergency set-down with engine trouble) and A-4044. During its delivery from Queenstown on 16 October, A-3478 attacked a suspected U-boat. The first bomb failed to detonate but the second exploded close to the periscope. Surface oil and debris was subsequently spotted, which the crew interpreted as serious damage at least.

That said, most patrols were without incident. On 14 October though, Queenstown's A-3468 bombed a suspicious-looking oil patch off Wexford harbour and four days later spotted a suspected submarine wake, bombing ahead of it but inconclusively. On 9 November, A-1048 patrolled for an in-theatre H-16 record of 9 hours 37 minutes.

Meanwhile, on 20 September, Lough Foyle's A-891/LF-3 had attempted to attack a suspect but its bombs wouldn't release, the mechanism having been set too tight. A-1032/LF-4, escorting a 32-ship convoy, bombed an incoming U-boat on 19 October; the submarine disappeared and again, oil surfaced at the spot.

A Wexford-based machine attacked a suspect in mid-September, while a similar incident on 11 October yielded an oil slick in the vicinity that lingered for a week. Eleven days later, Whiddy Island's A-1072 crashed while attempting a nocturnal landing; its wireless operator died, the only fatality among Ireland's H-16 crews.

Patrols ended on 11 November and training flights soon after. By the Armistice, only 43 machines had been delivered of which 28 had flown, Wexford making the most flights (98). By then, Queenstown's

H-16 A-1076 entering Queenstown's waters, assisted by handlers wearing waterproof waders. (Colin Owers)

Anonymous H-16 at Whiddy Island. Surrounding hills provided some shelter from the elements but could make local flying bumpy.

A battered H-16 on Queenstown's slipway. All wingtips in view are damaged, the port wing float lost.

personnel consisted of 48 officers and nearly 1,500 men, although none of the stations had achieved their planned complements or operational targets; there hadn't been time to fully work up. But despite their engine troubles, the H-16s had flown a total of 761 hours 24 minutes, sometimes in poor weather. And for two people in particular, events had worked out very well: during his stay Cmdr McCrary married US Navy Chief Yeoman Mary Davison in Dublin.

Lough Foyle, late summer 1918: ground crew with an H-16, wooden scaffolds beneath its Liberty engines.

Lough Foyle: H-16 BuNo A-779 LF-5 at rest on its beaching chassis around the time of the Armistice. To the left, the station's hangars.

Early spring 1918: Wexford under construction, its hangars partly built, the slipway and hutted accommodation in place.

H-16 station identity W3 on Wexford's apron. It sports a spigot mounting for a bow machine gun, but few of the Irish-based H-16s seem to have carried such weapons.

Anonymous H-16 on its beaching chassis at Wexford with air and ground crew, the Belfast hangars behind.

Wexford's hard standing: naval personnel with an H-16, a flyer to the left, ratings to the right.

Part 2

War of Independence, 1919–1921

Background

After the Great War, Ireland's military aviation presence quickly reduced. Airship flights ended in December 1918. Throughout 1919, the TDSs were disbanded but their airfields retained (though Tallaght was relinquished in March 1920). By April 1919, America's bases had been dismantled, occupied land returned, surviving aeroplanes and useful equipment shipped home. Rathmullan and Berehaven disbanded in March and August 1919 respectively.

But from the beginning of 1919, Ireland's domestic situation had become incendiary. On 21 January, the same day as Dáil Éireann's declaration of self-government, men from the newly formed nationalist Irish Republican Army (IRA) killed two Royal Irish Constabulary (RIC) officers at Soloheadbeg, Co Tipperary. This incident is widely seen as sparking the Irish War of Independence, a guerrilla (and sectarian) campaign fought between the IRA and the Crown. Today, the nationalist force is often referred to as the Old IRA, to distinguish it from later organisations bearing the same name.

Baldonnel nearing completion, January 1919, with Bessonneau hangars and permanent Belfast Truss counterparts. Between the two types sits an Avro 504.

F.2B F4796 of 106 Squadron under power at Collinstown, 1919. (J M Bruce/G S Leslie Collection, via Stuart Leslie)

Later in January, *The Aeroplane* magazine led with a curious, patronising piece titled, 'On the Irish Republic and Aeronautics', asking 'Who can deny Ireland's right to be a republic?' But it commented too, of the RAF stations there, 'Republic or no republic, these bases are an Imperial necessity', which rather missed the point. The nationalists' new publication *An t-Óglác* (*The Volunteer*) declared that a state of war existed between Ireland and Britain.

The IRA embarked on raids for arms and funds. Ambushes multiplied, mail trains were rifled, administration and army personnel killed. Hundreds of RIC officers – the British government's eyes and ears in Ireland – were slain, RIC barracks and loyalist properties destroyed. Hunger strikes began, while

Wg Cdr Keevor in 117 Squadron's DH.9 C1357 A, Fermoy, 1919. (J M Bruce/G S Leslie Collection, via Stuart Leslie)

for seven months the Irish Transport and General Workers' Union declined to carry service personnel by train. Dublin dockers refused to handle the Crown's military materiel, including aircraft spares and fuel. Graffiti daubed on street walls included 'Join the RAF and see the world, join the RIC and see the next world.'

For their part, the British employed increasingly brutal policing and use of armed force, committing raids on homes and businesses, lootings, arrests of nationalist suspects, beatings and extra-judicial killings. To bolster the RIC, notorious security contingents the Black and Tans and the Auxiliary Division were created. The army's Irish garrison grew from 53,000 troops in May 1919 to 80,000 in July 1921, conducting general military activities but also supporting the civil powers. Martial law was declared in counties Cork, Kerry, Limerick and Tipperary. Executions began. Between January and July 1921 alone, around 1,000 people died: RIC, army, IRA, and civilians on both sides.

The RAF Campaign

During the war, as best it could, the RAF undertook a co-operation role with the army and security forces. Aerial reconnaissances searched for IRA ambushes and drilling locations. Checks were made for sabotaged bridges and railway lines, cover provided for military motor convoys, freight- and troop-trains. Stolen vehicles – potentially, improvised armoured cars – were sought. The aircraft dropped messages to Crown forces, and leaflets to the population with details of wanted nationalists. In some areas, military documents, mail and VIPs were transported by air in relative safety compared with land movements.

Initially deployed were the resident Bristol F.2Bs of re-equipped 105 Squadron (relocated to Oranmore and Castlebar) and 106 Squadron (Fermoy). In March 1919, 141 Squadron moved to Tallaght (later Baldonnel) also with F.2Bs. That month, under-strength 117 Squadron (DH.9s) transferred from Wyton to Tallaght (later Gormanston) while 149 Squadron also arrived there, though at cadre strength. By that time, rather than being shipped to Ireland, the aircraft flew across from Britain, their dispersal point Baldonnel.

DH.9 D9883 of 22 TDS at Gormanston, early 1919. (J M Bruce/G S Leslie Collection, via Stuart Leslie)

F.2B H1488 of 106 Squadron at rest at Fermoy, 1919. (J M Bruce/G S Leslie Collection, via Stuart Leslie)

Gormanston, early 1919, at that time home to 22 TDS. Its coupled and single Belfast Truss hangars are supplemented by a Bessonneau.

The five units were merged into two: reformed 2 Squadron (F.2Bs), and 100 Squadron (F.2Bs and DH.9As, later just F.2Bs). Both went operational in February 1920 under No.11 (Irish) Group – amended to Wing status two months later – which itself had reformed on 22 August 1918, headquartered at Baldonnel under Gp Capt Ian Bonham-Carter.

Based at Oranmore, Castlebar and Fermoy, 2 Squadron's three flights positioned to other airfields as required. The four flights of 100 Squadron operated from Baldonnel and Castlebar but from time to time too, moved elsewhere. In November, 4 Squadron's A Flight travelled from Farnborough to Aldergrove (later Baldonnel), again with F.2Bs. In March 1921, British prime minister David Lloyd George authorised Irish Wing machines to carry machine-guns, but no records indicate their having fired shots during the war.

✠ ✠ ✠

The RAF struggled. Michael Brennan (later Lt Gen Brennan), chief of the IRA's East Clare brigade, commented after the war that with more aeroplanes to detect them, the British 'would have made short work of us'. But repeatedly, flying was hindered by shortages of fuel, spares and ground crew. Aging aircraft were patched up rather than properly overhauled; serviceability slumped. In March 1921, 100 Squadron's diary noted that three patrols supporting the army in 'rounding up [the] IRA' near Ballinrobe, Co Mayo, had been abandoned due to engine troubles.

Airfield accommodation could be depressingly poor; a report described Fermoy as 'squalid to the last degree'. At times, Bessonneau hangars were found unequal to Ireland's climate. Ground communications

Tallaght air and ground crew line-up along with an Avro 504J, 1919.

Sandwiched between a Bessonneau hangar and a more permanent type, a visiting 105 Squadron F.2B at Gormanston during 1919. (J M Bruce/G S Leslie Collection, via Stuart Leslie)

Very few SE.5as served in Ireland, but E5799 operated from Collinstown. (J M Bruce/G S Leslie Collection, via Stuart Leslie)

weren't always satisfactory; although Castlebar had wireless, there was no telephone until late 1921, the local RIC relaying messages.

Despite growing use of perimeter barbed-wire entanglements, and trip-wires connected to Verey lights or Mills bombs, the airfields became nationalist targets. In March 1919, the IRA attacked Collinstown, seizing 75 rifles and 5,000 rounds of ammunition, and damaging vehicles. June 1919 saw Ballyquirk's unfinished station raided for arms, though unsuccessfully. Periodically Baldonnel drew sniper fire and in August 1920, the IRA broke in, removing military ciphers. A machine-gun attack struck at Fermoy in May 1921, but without casualties. In November 1922, a mission to capture Baldonnel was reluctantly abandoned for lack of men.

Right: In February 1920, Gp Capt Ian Bonham-Carter became OC No.11 (Irish) Group, remaining in command when the force reformed two years later as RAF Ireland. In 1925, he was promoted to A/Cdre.

Below: Flamboyant F.2B F4380 of 105 Squadron, Fermoy, 1920. Behind is Avro 504K E403. (J M Bruce/G S Leslie Collection, via Stuart Leslie)

Above: C Flight, 105 Squadron at Castlebar. Bare-headed, CO Capt William Urquhart Dykes sits in the middle row.

Left: F.2B F4351 and another of 105 Squadron at Castlebar during the War of Independence.

Below: Baldonnel, 1920: 100 Squadron Handley Page O/400 J2259. On 17 December 1920, while flying from Britain to Baldonnel, J2259 crashed in the Irish Sea. The five crewmen and two passengers, one an Royal Irish Constabulary (RIC) constable, the other possibly from Britain's security services, were rescued by the steamer *Itajahy*. (J M Bruce/G S Leslie Collection, via Stuart Leslie)

In its 20–27 March 1921 edition, Milanese weekly newspaper *La Domenica del Corriere* ran a cover image of RAF aircraft strafing a handful of IRA men by a cottage. Its caption read: 'The airmen foil an ambush by rebels on [British] trucks loaded with troops, killing five of the assailants.' The episode was wholly imaginary.

Accidents and Losses

Various ordeals befell the RAF's aeroplanes, chiefly due to accidents but sometimes owing to the IRA's work. Avro 504Js C6018 and C5895 – both Harland & Wolff-built – collided over Collinstown on 13 January 1919; C6018's occupants perished. On 9 March 1919, 106 Squadron F.2B F4352 struck a cow while landing near Skibereen, Co Cork; damaged, the aircraft pulled up in a bog. Performing aerobatics at Birr on 28 March, the pilot of Avro 504J C5970 (again from Harland &Wolff) misjudged his loop. Clipping some trees, the machine hit nearby Crinkill House; the pilot survived but his passenger died.

Travelling from Gormanston to Aldergrove on 21 May, 22 TDS's DH.9 D9856 crashed at Rahanna, Co Louth, with engine problems. Though it was wrecked, no-one was hurt. F.2B F4669 of 105 Squadron spun in at Castlebar on 5 June; the pilot died, his observer survived. Avro 504K E326 crashed in the sea off Skerries, around 20 miles north of the capital, on 17 September, but the crew were saved.

At Drominagh near Clonbanin, Co Cork, on 14 August 1920, a 2 Squadron F.2B made an emergency landing with engine problems. It was guarded but the IRA arrived, killing a 17-year-old soldier before being forced to retreat. A week later F.2B H1571 came down near Lismore, Co Waterford. The IRA attacked again, that time burning the machine. The Crown issued a Malicious Injury claim seeking reparations from local residents (one of numerous such demands), but it's unlikely this was paid.

On 28 August, Maj Henry Chads of The Border Regiment died when 100 Squadron's F.2B F4528 crashed at Castlebar. His pilot, F/O Norman Dimmock, was seriously injured. Dimmock told the subsequent enquiry he'd been approaching too fast, intending to abort the landing and go around again. However, the machine struck the airfield's perimeter fence, collecting barbed-wire in its propeller as it came down.

At Fermoy, on 1 November, a 2 Squadron F.2B collided with another wandering cow. Also of 2 Squadron, F.2B H1487 crashed at Thomastown, Kilfinnane, Co Limerick, on 10 February 1921 and was burned by the IRA's East Limerick flying column. F/O N V Moreton, H1487's pilot, evaded capture but his observer, F/O E F MacKay, was seized. The RAF threatened to bomb Kilfinnane if MacKay wasn't safely returned; as a show of strength F.2Bs flew over the town. Instead of bombs, smoke-markers were

Above left: On 21 February 1920, South African pilot Lt Croye Rothes Pithey led a ferry flight of three F.2Bs (H1567, 2 Squadron; H1621, 2 Squadron, F/O H de W Waller; and H1612, 100 Squadron, F/O H L Holland) from Shotwick to Baldonnel. Holland had made the trip previously. The weather was reportedly favourable on both sides of the Irish Sea but still the machines disappeared; no-one was ever found.

Above right: During 1921, F.2B H1590 of 2 Squadron came to grief at Fermoy. (J M Bruce/G S Leslie Collection, via Stuart Leslie)

Above left: On 17 November 1920, F.2B H1490 with Cyril Ferdinand, F/Os Briggs and F W McKiehan (sometimes McKeechan), carrying despatches from Fermoy to Waterford barracks, crashed on to Mr Aspel's licensed premises at 39 Barrack Street, adjacent to their destination. The crew were taken to hospital in Fermoy and the Bristol repaired.

Above right: F.2B identity B in Ireland at the time of the War of Independence, its skull and crossbones motif suggesting a 100 Squadron machine. (J M Bruce/G S Leslie Collection, via Stuart Leslie)

Below: Returning from a reconnaissance F.2B J6696 of 2 Squadron suffered an accident while landing, nosing in by a haystack.

Baldonnel: F.2B engines from scrapped machines, soon to be further reduced.

dropped but MacKay was duly released. Down the years, H1487's propeller survived; today it's with Dublin's National Museum of Ireland at Collins Barracks.

On 20 May, a 2 Squadron F.2B force-landed outside Limerick, its crew setting fire to their despatches and then leaving to get help. But IRA men, hurrying to the site, found the documents – containing information on their activities – largely undamaged and took them, though on that occasion there wasn't time to destroy the aeroplane.

Truce and Treaty

In July 1921, with the conflict at stalemate a truce was called. Aerial activity reduced. From October, talks in London between nationalist representatives and the British government sought to agree the form of a treaty whereby the introduction and extent of self-rule for Ireland would finally be settled.

The nationalist negotiators included Michael Collins, a man of many aspects: a Sinn Féin Westminster MP, finance minister for Dáil Éireann, the IRA's director of intelligence, and organiser of a group of assassins, with a British price on his head (suspended for the talks). Fearing Collins' detention should dialogue fail, the nationalist government-in-waiting instructed two IRA men to buy an aeroplane as a means of his escape from London, if necessary. Charles Russell and William Jasper (Jack) McSweeney were both former RAF flyers.

Acting undercover, Russell bought a Martinsyde Type A Mk.II (c/n 217) from the Aircraft Disposal Company (ADC), modified to carry five passengers. He felt that should war resume, as well as evacuating Collins, the machine might be adapted for reconnaissance work, or bombing missions in Ireland or even Britain. For future pilot training and for dropping small bombs if required, Russell also purchased an Avro 504K from ADC.

The Martinsyde moved to Brooklands from where, should the need arise, it was planned to fly Collins direct to Dublin's Leopardstown racecourse. In London, meanwhile, British pressure to accept their conditions or face renewed hostilities added to the negotiations' fraught atmosphere. Finally, on 6 December 1921 the Anglo-Irish Treaty was signed, the emerging Irish Free State's form agreed in principle. Collins, though not wholly satisfied with the settlement – which conferred autonomy, but by no means full sovereignty – believed it provided the best possible stepping-stone to securing an entirely independent (and ultimately, fully united) nation in the future.

After passionate and acrimonious debate, the following month the Dáil voted 64–57 to accept the treaty, the terms of which included self-governing Dominion status for the new country as part of the

Top: DH.9A H3586 of 100 Squadron, Fermoy, 1921. By the tyre is a sack containing Crown mail and documents. (J M Bruce/G S Leslie Collection, via Stuart Leslie)

Above: F.2Bs of 2 Squadron prepare to start up at Baldonnel.

Right: During the War of Independence Curragh Camp became home to F.2Bs.

Commonwealth, an oath of allegiance to the Crown, and partition of Northern Ireland. Headed by co-negotiator Arthur Griffith and Collins, a provisional government was formed, charged with transfer of power from Britain to the Free State. However, the treaty seismically split the Dáil, Sinn Féin and the IRA. For the incensed anti-treaty faction seeking complete independence, led by former Dáil president Éamon de Valera, its terms were intolerable; if liberty was not entire, it was not liberty.

DH.9A H3553 of 100 Squadron outside a Baldonnel hangar, together with a Hucks starter.

Left: **F.2Bs of 100 Squadron at Baldonnel, early 1922. The unit was about to return to Britain.**

Below: **F.2B running up, its location probably Curragh Camp but possibly Oranmore.**

Part 3
Civil War, 1922–1923

Outbreak
Following the treaty's signature, British forces began leaving Ireland. In February 1922, 2 Squadron moved to Digby and 100 Squadron to Spittlegate. Travelling in the opposite direction, on 16 June, Russell's crated Martinsyde finally arrived at Baldonnel. Painted silver, Free State tricolour markings in green, white and orange were applied. The machine received no civil or military identity, though the engine cowling bore its new name, *The Big Fella*, which was also Michael Collins' moniker. The Avro 504K arrived the same month.

But the internal strife over the treaty plunged Ireland into a further period of upheaval, the Dáil's profound divisions reflected in the country at large. March 1922 saw a stand-off between hundreds of armed pro- and anti-treaty supporters in Limerick. That month, too, the British ship *Upnor*, leaving

Incoming: the Air Service's first aeroplane, Martinsyde Type A.II *The Big Fella*, Baldonnel, October 1922. A proud ground crewman applies the new national marking.

Outgoing: F.2Bs and personnel of the RAF's Irish Flight at Collinstown, 7 October 1922, shortly before leaving for the north.

Cork for Devonport with arms and ammunition, was captured by anti-treaty IRA men. In April, pro-treaty Gen George Adamson was killed by anti-treatyites. May saw severe clashes in Kilkenny city, 18 people dying in the fighting. Finally, during June, full-scale civil war erupted between the provisional government, deploying troops drawn from the pro-treaty IRA element – which became the state's National Army – and the faction backed by de Valera, whose forces were anti-treaty IRA fighters often known as the Irregulars.

In the meantime, during February, No.11 (Irish) Wing had been renamed RAF Ireland, still under Bonham-Carter. Two months later, to help protect the departing British army from possible nationalist attacks, an Irish Flight formed at Baldonnel with an average presence of four F.2Bs. From May, the flight was based at Collinstown, asked to vacate Baldonnel by the provisional government who wanted it turned into a civil airport.

In June, to further reinforce the army's safety, Aldergrove in the north – not fully operational but retained by the RAF for occasional use – received 2 Squadron. That month, Britain's remaining aerial assets in Ireland passed to No.12 Wing. In September, 2 Squadron left again; the following month the Irish Flight moved to Aldergrove, disbanding in November. With the army gone, in February 1923, No.12 Wing (by then based at Aldergrove) also disbanded.

Fledgling Air Arm

Although finances were extremely tight, Michael Collins was keen to develop aviation in the new country. On 23 March 1922, Charles Russell and Jack McSweeney became Free State director of civil aviation and director of military aviation respectively. That said, their departments were tiny; by July, just 16 people staffed the civil aviation office and 36 the military arm. The civil side was based at Baldonnel, the military at Dublin's Beggars Bush army barracks before moving to the airfield. McSweeney's group became known by several titles, including the Army Air Service and the Military Air Service. In general use, this was often shortened simply to the Air Service.

With political and social conflict raging, by July the civil department had been subsumed within the Air Service, led by McSweeney with Russell as his deputy and forming the aerial component of the National Army. On 1 July, Collins became Minister of Defence, and from 12 July, commander-in-chief of the armed forces. He appreciated aeroplanes' value for reconnaissance work, and for helping protect potential targets from attack.

Meanwhile, in an act of symbolic confrontation also intended to provide accommodation for their fighters in the capital, during April, Irregular forces had seized Dublin's Four Courts, the country's most prominent judiciary buildings. Unopposed until 28 June, finally troops began shelling the site. Winston Churchill, head of Britain's Colonial Office and chairman of the Committee of Imperial Defence sub-committee on Ireland, proposed the RAF should lend aircraft to the Free State, flown by RAF personnel but wearing Irish markings. With attacks on the Four Courts in mind, the Irish Flight began bombing practice. F.2B H1485 was earmarked but the plan wasn't taken up, partly because Collins feared civilian casualties, and by 30 June the Four Courts' occupiers had surrendered.

However, following an appeal by Collins to the Colonial Office, two F.2Bs passed to Free State ownership. Collins requested aircraft on 4 July after widespread acts of sabotage by the Irregulars, ground searches for whom had failed. The next day F.2B E2411 was collected by Russell. Folklore tells us that for the hop from Collinstown to Baldonnel he wore a bowler hat. The machine received Free State identity Roman I – soon changed to B.I, the B for Bristol, then BF.I (BF: Bristol Fighter) – and national tricolour markings were added.

Above left: William McSweeney (left) and Charles Russell, co-architects of the Collins escape plan and the Air Service's first pilots.

Above right: Michael Collins led the nationalist representatives who negotiated the Anglo-Irish Treaty and in July 1922 became commander-in-chief of the Free State's armed forces, including the Air Service.

Above: Avro 504K A.I at Baldonnel, October 1922. In its rear seat is Lt William Delamere. The man by the wing appears to be praying.

Left: Tail-up at Baldonnel, khaki-painted Avro 504K A.III of A Flt, No.1 Squadron.

Below: Limerick's Fair Green, August 1922. F.2B B.I came down after experiencing engine trouble caused by inferior fuel. Troops guard the aircraft, and two ladies hold rifles while their photograph is taken. B.I has no visible markings.

By then too, McSweeney had been instructed urgently to acquire another aeroplane. From ADC he bought F.2B H1251, which was test-flown on 3 July. He left Croydon the following day and via RAF Shotwick arrived at Baldonnel on 6 July. Initially the aircraft acquired identity II, later B.II and finally BF.II.

The second ex-RAF machine, H1485, appeared on 10 July, becoming III (later B.III). But the new force faced severe shortages of pilots, and repairs and maintenance facilities. Some materiel was extracted from the departing British, while slowly a team of fitters and riggers was assembled.

February 1923: B.I at Baldonnel bearing national and identity markings, Lt Delamere toting a rifle. A year later the machine crashed and was written off.

Khaki-painted F.2B being pushed into a Baldonnel hangar. It wears national markings on the tail and fuselage, but has no visible identity.

The Air Service Campaign

Though McSweeney headed the Air Service, until August Collins directed its operations. The first recorded reconnaissance took place on 16 July 1922, with McSweeney and observer Lt Thomas Nolan in F.2B III. From Baldonnel, they flew south-west, but apart from roadblocks outside Baltinglass (Co Wicklow) and Tallow (Co Waterford) spotted nothing untoward. However, during their return, the machine experienced fuel-flow problems; McSweeney force-landed at Ballycane, Co Kildare. Nolan was injured in the crash, while the Bristol stayed non-operational until February 1923.

The following day, Russell, Staff Capt W Stapleton and (probably) F.2B I made another flight. Over Baltinglass the aircraft acquired several bullet holes, but low mist prevented Russell from strafing his Irregular assailants. He reported opposition as appearing generally light, and later that day the town was occupied by Free State troops.

During July, as the Irregulars retreated, altogether around a dozen flights scouted south-west of Dublin, encountering occasional ground fire. Over Co Kilkenny's towns on 22 July, Russell and Stapleton dropped bundles of 'war special' copies of *An t-Óglác*, by then a Free State newspaper. The F.2B's endurance limited patrols from Baldonnel to parts of Co Cork, so at Collins' behest two landing grounds were prepared at Kilkenny and Waterford, the latter by the racecourse. By early August, both were operational.

On 7 August, Russell and F.2B I set off to survey the Cork area, drop a special 'air' edition of *An t-Óglác*, and provide air support, if required. Alighting at Kilkenny and Waterford, he was delayed by bad weather but on (probably) 9 August made his reconnaissance. Passing over Cork city he noted fires and smoke, and crowds of men. Irregular forces didn't wear uniform so he was unsure who the men were, but the area was an anti-Treaty stronghold. Russell's flight back to Waterford was interrupted by ground fire, which he returned. The mission coincided with Free State troops' attack on the city, which fell on 10 August.

By mid-August, Russell and F.2B I were flying from Limerick's racecourse and Fair Green, patrols covering Cork and Tipperary. Various sabotaged road and railway bridges were spotted and occasional

Early 1923: pilot Lt James Fitzmaurice and observer Lt L C Gogan in F.2B B.V, by Fermoy's skeletal hangar A. On 6 August 1924, B.V crashed at Crumlin, Co Dublin, following engine failure and was written off.

Civil War, 1922–1923

Right: F.2B under maintenance at Baldonnel, its cowlings removed. Ground crew pose for the camera.

Below: Avro 504K A.III and servicemen at Baldonnel. The machine endured until 1931.

shots hit the machine. Meanwhile, two new pilots had appeared: Lts Fred Crossley and Thomas Maloney, both ex-RAF.

McSweeney bought more aircraft from ADC: a Martinsyde F.4 Buzzard and an SE.5a. Delayed by bad weather and mechanical problems, the Martinsyde (identity I, later M.I: Martinsyde, and christened *The Humming Bird*) arrived at Baldonnel on 15 August, the SE.5a following and given identity II.

In August too, Collins, McSweeney and Russell met, agreeing that to allow reconnaissances further south-west over the Irregulars' heartland, another new airfield would be needed. Fermoy's ex-RAF station

Above: F.2B at Baldonnel. This aircraft served as VII, BF.VII and finally 7, until struck off in April 1934.

Left: Baldonnel: Free State aircrew pose for their photo – the pilot rather bashfully – in an F.2B, the observer holding his Lewis gun.

north of Cork city was agreed on. This was their last meeting; later that month, *An t-Óglác* reported the death of Michael Collins, ambushed and shot by Irregulars on 22 August at Béal na mBláth, Co Cork.

✠ ✠ ✠

The SE.5a wore an unusual colour scheme: a red fuselage, silver wings (their old RAF roundels overpainted with Free State colours), its rudder carrying the national tricolour. But on 8 September, Fred Crossley was flying to Limerick when visibility reduced. His compass became faulty and the engine developed problems; lost, he force-landed near Macroom, Co Cork. Crossley survived but the SE.5a, its weapons removed, was burned by the Irregulars.

That month Gen Richard Mulcahy succeeded Collins, while four more pilots arrived. McSweeney bought further machines: five F.2Bs, three Buzzards and three 504Ks. The first F.2B, IV, fitted with racks for 20lb bombs, touched down at Baldonnel on 16 September. Damaged while landing, it never served; its engine later passed to B.II. Deliveries continued over the next two months, the F.4s appearing on 14 October. The aircraft joined the Air Service's recently commissioned No.1 Squadron, trainers with A Flight and fighting types with B Flight.

On 1 October, Fermoy hosted its first Free State aircraft: Buzzard I flown by Lt James Fitzmaurice, again ex-RAF, who'd joined the Air Service in February. Crossley and F.2B B.II also arrived. It hadn't been easy to revive Fermoy; much of the equipment and structures left by the RAF had been auctioned off or pinched, including the hangars' corrugated iron cladding.

Initially the aircraft lived under tarpaulins but some cladding was retrieved from surrounding farms and one of the three hangars restored. Huts arrived for personnel and storage of munitions. Again, missions involved reconnaissance, protection of roads and railway lines, distribution of *An t-Óglác* and cover for advancing Free State forces.

Anonymous Martinsyde F.4 in a Baldonnel hangar, wearing national insignia on its rudder. No other markings are in view.

At Baldonnel a Martinsyde F.4 displays bomb racks just aft of the main undercarriage, so this is possibly identity III.

Sometimes carrying a rifle while flying, Fitzmaurice had numerous adventures. In October, the government offered a pardon to Irregulars who surrendered their arms, and accordingly he was employed in scattering circulars to that effect over Cork and Kerry. But on 14 October, near Killarney, his Buzzard's Hispano failed. Alighting by the Killarney District Lunatic Asylum, he worked on the engine until the following evening; while taking off, gunfire hit the Martinsyde's wings and fuselage. Via another forced landing near Mallow, Co Cork, Fitzmaurice returned to Fermoy by road. The machine was towed there for repairs, flying again on 8 December.

That month too, the *Irish Times* reported an engagement near Dunmanway, West Cork.

> About 60 armed men sighted the military in two lorries and prepared to ambush them… the troops realised their attackers were in excellent defensive positions and sent for reinforcements. An aeroplane also came. It was of the small scouting type. The pilot… when just over the ambush party nose-dived in their direction. Panic at once seized them… bombs were dropped from the aeroplane and the party was filled with consternation.

The aircraft was most likely a Buzzard fitted with an under-fuselage bomb rack. By then, the three later Martinsydes had received identities II, III and IV (later M.II, M.III and M.IV) and III at least was so-equipped. The Irregulars ran towards a wood, the Buzzard machine-gunning the area before leaving. In the action several people died.

Amidst the conflict, with the Anglo-Irish Treaty's ratification in December, the Free State officially came into being. That month too, a final wartime airfield was created near the militia base and racecourse at Cloon More, Tralee, Co Kerry. At first, its force generally consisted of just one F.2B (at least B.I and B.VIII at different times) usually flown by new pilot Lt William Delamare. As Free State troops continued pushing south-west, his main task was reconnaissance over Kerry.

✣ ✣ ✣

By January 1923, the Air Service had ten pilots. Subject to weather, operations continued through the winter and spring. On 16 January, Delamare drew ground fire near Killarney, in return releasing two

Above left: James Fitzmaurice became one of Ireland's most prominent aviators. In April 1928, he co-piloted Junkers W.33 D-1167 *Bremen* on the first-ever non-stop transatlantic aircraft flight from east to west.

Above right: Engineless Martinsyde F.4 M.I *The Humming Bird* at Fermoy. In the foreground, a second example.

bombs and machine-gunning the Irregular force until Free State soldiers arrived. Over January and February, six newly-purchased ex-ADC DH.9s (initially I–VI, later D.I–D.VI: D for de Havilland) had joined No.1 Squadron's B Flight. Two served with Fermoy's detachment, again undertaking reconnaissance, communications protection and troop support.

One day early that year, promoted to captain Fitzmaurice, he made a flight to Co Cork, to find and bomb a Rolls-Royce armoured car seized by the Irregulars. The vehicle wasn't located but another forced landing ensued, near Kinsale Junction. Hostile people gathered; Fitzmaurice created a retreat by telling them that at any time the Buzzard might explode. Purloining a horse, he rode bare-back to the nearest government barracks for help. Meanwhile, overcoming its fear, the crowd attempted to burn the machine but arriving soldiers mounted a guard. The dismantled Buzzard was transported safely back to Fermoy.

During March, Delamare was busy dropping leaflets encouraging the Irregulars to surrender. By June, six 504Ks (I–VI, later A.I–A.VI: A for Avro) were on strength, including the machine bought along with *The Big Fella*. Identity IV was Harland & Wolff's ex-E359, the only Air Service aeroplane built on the island of Ireland.

✈ ✈ ✈

On 24 May 1923, the beleaguered Irregulars finally laid down their arms. Over the summer though, aerial patrols continued, to confirm peace was being maintained. On 25 June, DH.9 D.I, seen spinning over Fermoy, crashed nearby; observer 2nd Lt John McDonagh died, the only Air Service fatality of the war and its aftermath. In October, Tralee was vacated, and Fermoy in April 1924, although some private flying took place there over later years.

What of *The Big Fella*? Despite McSweeney's earlier optimism, it wasn't ideal for reconnaissance or easily adapted for bombing. An engine transplant helped delay its first Irish flight until 13 October 1922. Post-war it provided occasional VIP transport and was flown by Fitzmaurice, who in 1927 became Baldonnel's commandant. By then re-named *City of Dublin* (on the engine cowling port side) and *Catair Ata Cliat* (on the starboard side) it was retired that year, becoming an instructional aid at Baldonnel.

Left: Three Free State DH.9s, Baldonnel, 1923. The lead machine wears underwing tricolour markings and tail insignia.

Below: *The Big Fella* at Baldonnel prior to a flight before *An t*-Óglác representatives, 10 February 1923. Its name has been applied to the engine cowling, and tribands to the rudder as well as below the cockpit.

Bottom: Air Service line-up at Fermoy, 1923. Left to right: Martinsyde F.4, F.2B BF.VII and two DH.9s.

After the war the Air Service came under great scrutiny, not least for financial reasons, and the need was questioned for any military aircraft at all. Mulcahy was much less committed to aviation than Collins had been. However, the force survived and a reorganisation was approved by the Executive Council (the *de facto* executive branch of the government), whereby under Orders No.3, Defence Forces (Organisation) Order of 1 October 1924, the Irish Air Corps was created, continuing as part of the army and Baldonnel remaining its main airfield.

In 1925, six more F.2Bs arrived, partly as what we'd today call attrition replacements. By then, the Air Corps had introduced Arabic numerals to identify its aircraft (though with a couple of exceptions, not retrospectively), the new F.2Bs assuming 17–23. These were Mk.II variants, 18, 19 and 22 serving until June 1935. Meanwhile, the final surviving Buzzard, M.I, had crashed near Baldonnel in May 1929 and was written off; pilot Lt Arthur Russell (Charles' brother) was seriously injured. DH.9 D.III endured until September 1934.

Right: Two DH.9s and a Martinsyde F.4, Baldonnel, 1923. On the right, James Fitzmaurice.

Below: Minus wings, tailskid and rear armament, red-nosed DH.9 D.II, its fuselage structure partly exposed. D.II was finally withdrawn in February 1930.

Above left: Michael Collins' successor, General Richard Mulcahy, wasn't as enthusiastic about the idea of an air corps; post-civil war, he pondered its continued need.

Above right: At the Air Service's first public display, held at Dublin's Phoenix Park in August 1924, pilot Lt Timothy Prenderville is congratulated by Maj Gen Paddy Daly. Behind the men, an Avro 504K.

DH.9 D.II and cheery crew during the August 1924 Phoenix Park display. The machine carries forward armament but lacks a rear Lewis gun.

Civil War, 1922–1923

Sporting a silver scheme, which replaced khaki for the type, an F.2B Mk.II passes over the ploughed Irish countryside.

Fitzmaurice in silver-painted Buzzard M.I during an army exercise. He flew with the army's 'Red Force', the machine accordingly wearing a red fuselage band.

On 21 September 1926, F.2B Mk.II 17 of B Flt, No.1 Squadron crashed at Hempstown, Co Wicklow, during army manoeuvres. Pilot Lt Timothy Prenderville and observer Lt Edmond O'Reilly perished.

F.2B Mk.II 18 at Baldonnel. Following a survey of obsolete aircraft and engines held in storage, 18 was finally scrapped in June 1935.

DH.9 7 was delivered in April 1929 as a replacement for struck-off examples, joining No.1 Squadron's B Flt and serving until September 1934. A second DH.9, which arrived at the same time, became 8.

F.2B Mk.II 19 with (left to right) Capt G Carroll, Lt McSweeney, Lt Russell, Capt Delamere and Lt Gogan, the machine having returned to Baldonnel from Oranmore after helping search for missing small boats.

Silver F.2B Mk.II 22 joined B Flt, No.1 Squadron, surviving until June 1935 when, along with 18 and 19, it was finally scrapped.

Dismantled and engineless, the Martinsyde Type A.II late in life. During the early 1930s, it was earmarked for display in Dublin's Mansion House, coincidentally the meeting place of the Dáil between 1919 and 1922. Instead, following instructional use and storage, the remains were scrapped in January 1937.

Part 4

The Emergency, 1939–1945

War Clouds All Around

By the mid-1930s, the Air Corps had purchased several more aircraft types, but in penny-packets. Post-civil war, the government and army had concluded threats to Ireland's stability were greater from persisting and at times serious anti-treaty IRA activity than from foreign powers, and to help maintain internal security, preferred the use of ground forces to aeroplanes. In turn, Air Corps funding remained scant.

Types acquired included the Fairey IIIF II, Vickers Vespa IV and V, and various Avros: the 621, 626, 631 Cadet, and 636. The 636s received Armstrong Siddeley Jaguar engines taken from struck-off Vespas. During the late 1930s, a few more modern machines arrived: Avro Ansons, Gloster Gladiators, Miles Magisters, Supermarine Walruses and Westland Lysanders. But a Vespa V, three 621s, two 626s, five Cadets and three 636s toiled into the 1940s.

From September 1936, the Corps' commanding officer had been Col Patrick Anthony Mulcahy, younger brother of Gen Richard Mulcahy. A former artillery staff officer with no aviation experience, just prior to his appointment's confirmation he began flying training. Although duly certified as a pilot, he didn't complete the syllabus.

Beyond Ireland, over the next three years, European tensions increased to breaking point. On the afternoon of 3 September 1939, RAF Saunders-Roe Lerwick L7252 encountered bad weather *en route* from Pembroke Dock to Stranraer. F/O David Banks spotted a gap in the clouds and descended, emerging south-east of Dublin over the port of Dún Laoghaire. Setting down, he presented himself to the harbourmaster. He was allowed to make purchases at a local shop and then sent on his way. Short Sunderland L2158, alighting at Skerries in the same murk, was also packed off.

The British crews and their flying-boats could have been detained; as war broke out, Ireland confirmed her neutral status and passed the Emergency Powers Act. This legislation included government regulation of the economy, food rationing, media censorship and internment. The period known in Ireland as the Emergency had begun.

Despite neutrality, aerial incursions or attacks were always possible, but the Air Corps was terribly ill-prepared to oppose any foreign aggression. Just three under-strength squadrons existed, plus the

Patrick Anthony Mulcahy with his first wife Josephine Barrett. Between June 1935 and December 1942, Col Mulcahy headed up the Air Corps.

Three of the first batch of Ansons at Baldonnel, wearing pre-Emergency colours: pea-green fuselage and engine nacelles, silver flight surfaces and polished cowlings. Centre is white-coded 21 and behind the Avros, DH.84 18.

Gladiators at Baldonnel in their pre-Emergency colour scheme, prior to October 1938: from the foreground, 24, 25 and 23. Behind, two Avro 636s.

Corps' Training Schools. Some 30 pilots were available, half the notional number required for even that force, while serious shortages also existed across the ground trades.

Formed in April 1937, No.1 Reconnaissance and Medium Bomber Squadron flew nine Anson Is, four new and five ex-RAF. The new batch had arrived between March 1937 and January 1938, becoming A19–A22 (post-1938, renumbered 19–22), the first monoplanes and the first aircraft with retractable undercarriage to join the Air Corps. The five later examples, delivered in February 1939, became 41–45. Seven more Ansons were ordered; ready for export by September 1939, with the outbreak of war they were embargoed by Britain. The squadron also operated a DH.84 Dragon II, identity DH18 (later 18). Airwork Ltd had converted it into a target tug and added small underwing bomb racks.

Gladiator 26's landing accident at Baldonnel, 2 June 1938. Its wing guns are absent.

Baldonnel: Walrus N20 shows off its lofty profile. In the background is a Magister.

Gladiators under maintenance at Baldonnel prior to the Emergency. For the camera, groundcrew scrutinise 24's main-wheel.

Saunders-Roe Lerwick L7252 at Dún Laoghaire, 3 September 1939. Its starboard engine is under power, the bow turret retracted to allow mooring.

With three Walrus Is, No.1 Coastal Patrol Squadron formed in May 1939, also acting as a training unit. Prior to export the amphibians received Supermarine B conditions identities N18, N19 and N20. During its delivery on 3 March 1939, N18 force-landed with engine trouble off Ballytrent, Co Wexford. Rescued by local boats, its upper wing damaged, the aircraft was trucked to Baldonnel. For some time, the B identities were retained but eventually the 'N' was dropped.

No.1 Fighter Squadron, formed in January 1939, flew three Gladiator Is (24–26), which had arrived in March 1938, and from July 1939, six Lysander Mk.IIs (61–66). Twelve more Gladiators had been ordered, but again were embargoed. In 1940, a squadron badge was devised, featuring a black panther's head, and later the motto (translated from the Irish language) 'Small but Fierce' added. The Lysanders performed mainly army co-operation duties, though Col Mulcahy felt they could act as fighters if need be. The three squadrons were often referred to without their 'No.1' prefixes; all were based at Baldonnel.

Ireland's fourth Gladiator, 23, never flew with Fighter Squadron but served only with a predecessor unit, No.1 Army Co-Operation Squadron. An accident near Baldonnel on 20 October 1938 following engine failure wrote it off after just 97 hours 40 minutes flying time, though pilot Lt Malachy Higgins escaped with minor injuries. On 2 June 1938, 26 had nosed over at Baldonnel with squadron CO Capt Michael Sheerin, staying inactive for two years. However, by summer 1940, all three remaining examples were serviceable.

Early Operations

Immediately prior to the Emergency, at short notice, three (later four) Ansons and two Walruses under Capt William Keane moved from Baldonnel to an isolated spot on the Shannon estuary, Co Clare. With a handful of pilots and ground personnel, coastal patrols began from an airfield recently created at Rineanna, near the Foynes transatlantic commercial flying-boat base.

Foynes, autumn 1939: Rineanna-based Anson 20 passes over a Pan American Airways Sikorsky S-42 flying-boat.

Rineanna had received its first visitor on 18 May 1939 – Anson 43 – but for some time, facilities were practically non-existent: no hangars, direction finding station or wireless. Maintenance beyond the basics and all repairs were undertaken at Baldonnel. A wireless car arrived to help out, and the nearby Ballygirreen civil aviation communications station agreed to provide a direction-finding service in between dealing with Foynes' traffic; fortunately there wasn't a great deal at that time.

Patrols, typically of three hours' duration, began on 31 August with 43, while three days later, Walrus N20 made the first such flight of the Emergency with Capt Finbar O'Cathain and W/Op Pte Connolly. Rineanna's aircraft, at first two each day, attempted to watch over a vast number of coves and small bays stretching from Donegal to Wexford, many of them isolated. It was surmised the inlets might be used by German craft as hideaways, while engaged in shipping attacks or espionage missions.

The patrols also recorded incoming weather patterns, sought damaged vessels and survivors of sinkings, and investigated reports of encroaching aircraft from what were termed in Ireland the 'belligerent nations'. On 4 September, an Anson left Rineanna to help spot survivors from the British liner *Athenia*, sunk off Donegal the previous evening by German submarine U-30. However, surface craft arrived to make a rescue and the Irish mission was curtailed.

Anson strength soon reduced sharply, three of the second-hand examples involved. On 8 September, 45 suffered engine failure following fuel starvation, being wrecked while force-landing at Ballyferriter near Dingle, Co Kerry, though no-one was seriously hurt. In heavy rain on 10 October, 44 came down and was damaged at Borrisokane near Nenagh, Co Tipperary. Lt P O'Sullivan's 43 crashed in Galway Bay on 19 December, after double engine failure with fuel problems off Barna, Co Galway. The crew was rescued, the aircraft towed into Barna but later struck off. Meanwhile, 44 was moved to Baldonnel but repairs took forever and it didn't fly again until June 1945.

Always unequal to covering such a large area, Rineanna's patrols declined to one a day. Spares arrivals were very slow (a situation that would endure throughout the Emergency) and by spring 1940, of the six remaining Ansons, three were unserviceable with engine defects. Flights became reactive, as and when Rineanna received notice of sightings it felt could reasonably be investigated.

However, coastal observation was bolstered by a national network of lookout posts run round-the-clock by the Marine and Coast Watching Service. This body had been formed in February 1939, after the Dáil had acknowledged its obligation to comply with the 1907 Hague Convention's provisions on the

Walrus N19 under power at Baldonnel, post-March 1939. Behind, Magister 39 gets away.

Walrus N20 over the sea off south-west Ireland. Its port wing national marking bears a large patch.

Borrisokane, Co Tipperary, October 1940: Anson 44 is recovered following its crash. More than 5½ years passed before it returned to service; in July 1946 it was withdrawn.

rights and duties of a neutral country in case of war. The posts, which grew to number 83, reported their sightings to army command in Dublin, from where the information was passed to the Air Corps. Some *Garda Síochána* (police service) stations also maintained watches.

By mid-year, with aerial patrols from the west suspended, the remaining Ansons had become Baldonnel-based. However, with the arrival at Rineanna of three Cadets – their silver-and-black scheme replaced by camouflage – that force resumed its work. Eventually Walrus 18 was repaired; by 1941 it was flying with the wings of 19, which had crash-landed at Baldonnel on 18 September 1940. On 3 September 1942, 20 was written off but the first example continued, although by then leaky wing floats prevented water-based operation.

Left: Ansons 20, 21 and 41 in Emergency camouflage. Also introduced at that time, the fuselage orange-and-green Celtic boss national marking was set within a white square and the numbering changed from white to black. Tribands were retained on the Ansons' upper wing surfaces until boss markings replaced them in 1944.

Below: Its identity painted out, camouflaged Walrus 20 off the south-west Irish coast. The new fuselage national marking lacks the square white background.

Avro Cadets, post-1938: left to right, examples 1, 2 and 7. All served during the Emergency, 7 at Rineanna with Coastal Patrol Squadron. Withdrawn in August 1945 after a ground accident, the machine survived and today wears C7, its original identity.

Baldonnel: Camouflaged Walruses 18 and 20 of Coastal Patrol Squadron, 1941 or 1942.

Incursions

The Dáil feared that despite neutrality, Ireland could become directly embroiled in the surrounding conflict. By 1940, the army had drawn up contingency plans in case the British attacked, either to occupy the country for their own security or to take control of strategic ports. Indeed, Britain ruminated on seizing the ports of Cork, Queenstown, Berehaven and Lough Swilly, viewing them as potential naval bases.

Equally, a landing by Germany along the south-east coast was believed a danger, and with their draft Unternehmen *Grün* (Operation *Green*) the Germans considered this. How serious were these ideas? Once Unternehmen *Seelöwe* (*Sealion*) was shelved, so too, with hindsight at least, were realistic prospects of *Grün*. Britain's thoughts also came to nothing. But at the time, for Ireland, the possibility of attack seemed all too real.

Over summer 1940, as Britain and Germany fought their terrible aerial dual, Dublin's airspace was declared a prohibited zone and belligerents' legations in Dublin notified that unauthorised aircraft entering the area would be fired on. Ireland's Air Defence Command was based at Dublin Castle, the capital's ground-to-air protection under No.1 Anti-Aircraft Brigade. By November, 14 anti-aircraft guns and a handful of searchlights had been assembled.

The warnings had little effect. Between May and July 1940, the look-out posts spotted an average of more than 400 foreign aircraft per month, particularly off the eastern coastline but also over Irish soil. An Air Corps interception procedure was drawn up, but this simply ignored the lack of resource. With so few aeroplanes, it was quite impossible for Fighter Squadron to tackle the numbers of intrusions reported, or organise standing patrols. In any case, Gladiators and Lysanders would have faired badly against belligerents' more potent machines.

Across Baldonnel, obstructions were prepared to hinder potential landings by intruders. With the possibility in mind of the base's evacuation during conflict, that summer numerous dispersed sites were identified as potential emergency airfields, if necessary, though fuel supply, maintenance and repairs at such strips would have been particularly difficult. The sites were mainly in Leinster and Munster provinces, some surveyed years earlier as potential venues for travelling aerial circuses. During the Emergency too, Fermoy accommodated Air Corps machines once again.

Autumn 1940: Lt Séan Kelleher, a graduate of 1939's Short Service Class completing on the Avro 636, sits in Gladiator 24 at Baldonnel.

Photographed from a Lysander, Emergency-camouflaged Gladiators 24, 25 and 26 over Wicklow's coastline, November 1940. Lt Séan Kelleher flew 24, accomplished aerobatic pilot Lt Des Johnston 26, and Lt M Maloney 25.

Summer 1940 also saw Fighter Squadron acquire newly camouflaged Avro 636s 14 and 17, the Dragon and a Magister, though this assortment hardly improved operational capability. On the prospect of pitting the unit's best aeroplanes against an attack on Dublin, Col Mulcahy commented: 'Supposing bombers came… and our three Gladiator pilots were shot down… it would be a certain consolation to the people and would improve their morale by letting them know that we had at least done what we could.'

After reconsideration of the overflight policy, for fear of breaching neutrality Fighter Squadron was ordered not to fire on belligerents' aircraft unless attacked. With the interception service plan unworkable, reports of single intruders became routinely disregarded. Facing no opposition, the *Luftwaffe* photographed various military bases including Baldonnel, but also the old Tallaght airfield site, which had never actually seen Air Corps usage.

✢ ✢ ✢

Parallel with planning against hypothetical invasion by her neighbour, and notwithstanding British ponderings on aggression, during 1940 dialogue with Britain spawned the possibility of in-country RAF support should Germany attack. Although the Air Corps had considered the possible evacuation of Baldonnel in such circumstances, it was mooted as a prospective RAF base. Gormanston, Collinstown and an advanced landing ground at Rosegarland near Wexford were also suggested, along with several of the emergency airfields then being contemplated.

Baldonnel: Lysander 61. Serving with Fighter Squadron and later the Training Schools, it was eventually scrapped in December 1946.

In the meantime, a few more aeroplanes arrived from Britain, but not modern fighting types. In June 1940, to help train sorely-needed new pilots, six Hawker Hinds appeared (67–72), though only three were dual-control. Five more Magisters (73–77) were also delivered, instead of nine promised Hinds. All joined the Training Schools, but on 27 July, Hind 70 crashed at Laytown, Co Meath; Lt Michael Ryan and Pte Patrick Power died. By September, just three Hinds remained, all unserviceable; one was repaired the following month.

Actions against belligerents' aircraft were few, sometimes muddled or missed altogether. On 13 April 1940, Air Corps machines investigated a large formation off the east coast, which turned out to be a British convoy escort flying beyond territorial waters. Irish-registered collier SS *Kerry Head* was sunk with all hands off Sheep's Head, Co Cork, on 22 October, the *Luftwaffe*'s attack unopposed. Two Gladiators sent to intercept a Junkers Ju 88 over Dublin in the mid-afternoon of 29 December failed to

Gladiators at Baldonnel; from the left, 25, 24 and 26. Below its cockpit, 25 sports a cartoon figure (possibly Sneezy of Seven Dwarfs fame) playing the concertina. Examples 24 and 25 had been camouflaged in March and May 1939 respectively, 26 in July 1940.

With a Gladiator forming the backdrop, Cmdt Daniel Horgan briefs aircrew at Baldonnel. Beneath the machine's cockpit the unit's badge is visible.

Lysanders at overcast Baldonnel, March 1941. This photograph was issued by G2 Branch (the army's intelligence service), rather optimistically captioned 'Light fighter-bombers tune up'.

Elegant black-and-silver Avro 636 fighter trainer A14 (post-1938, 14) joined the Training Schools in October 1935. From mid-1940, it flew with Fighter Squadron but was withdrawn in 1941. In fighter configuration the rear cockpit was faired over.

Lysanders exercising with a light motorised army column near Curragh Camp, summer 1940.

catch it, while scant anti-aircraft fire was also ineffective; the German machine simply rocked its wings and went on its way. March 1941 saw a Heinkel III cross the Co Wexford coast freely, before depositing spy Günther Schütz over Taghmon. The *Abwehr* man was later apprehended.

Successful offensive episodes were confined to the destruction of wayward barrage balloons. On 4 October 1939, Lysander 64 shot down a drifting balloon near Foynes, the remains of which helped improvise a hangar at Rineanna. On 20 June 1940, off Co Waterford, 64 dispatched its second balloon. The following May, a third was brought down near the county town of Carlow.

In May 1941, Britain delivered ten Hawker Hectors (78–87), and three more (88–90) the following January. Again, all joined the Training Schools. Obsolete, worn and employing the complex, troublesome Napier Dagger engine, understandably they were disliked. Mulcahy, disappointed, had been angling for Hurricanes, Harvards and Battles. In September 1941, Hectors 81 and 83 collided at Gormanston, being struck off. Remarkably though, 90 served until August 1945.

✢ ✢ ✢

Several times, German aircraft bombed Ireland. On 26 August 1940, two Heinkel IIIs crossed the Wexford coastline at Carnsore. One released four bombs over the village of Campile, killing three people. The other dropped a bomb on nearby Ambrosetown but no-one was hurt. Lt Andrew Wood, flying a scrambled Gladiator off the coast, noted visibility was 75 miles; he could pick out the Welsh coastline but the Heinkels had gone. There has since been speculation that rather than arising from error or accident, the act was deliberate. In fact, Germany admitted responsibility and apologised.

Hind 70 served for just two months before being written off in July 1940. Behind, a second example.

Two bombs struck near Dún Laoghaire on 20 December, injuring three people. New Year's Day 1941 saw the bombing of Julianstown and Duleek, both in Co Meath, but no-one was injured. On the night of 2/3 January, more bombs fell; three people died at Knockroe, near Borris, Co Carlow. Ballymurn, Co Wexford, Ballymany close to Newbridge, Co Kildare, nearby Curragh racecourse and Kilmacanogue, Co Wicklow, were also hit though without casualties, but in Dublin's South Circular Road area 22 people were injured by a single bomb.

On the starry night of 30/31 May 1941, the most severe tragedy struck Dublin's North Strand district. An estimated 30 bombers appeared; despite anti-aircraft fire their bombs killed at least 28 people (different sources say as many as 37), injured around 100 and destroyed or damaged 300 houses. On 2 June, Arklow, Co Wicklow, was struck and on 24 July Dundalk, but again without fatalities. Germany apologised for the North Strand bombing, promising reparation, but this was delayed until West Germany honoured the pledge in 1958.

After further talks with Britain over 1940 and 1941, and as discreet Irish assistance grew against the backdrop of a shifting world stage, fear of an attack from that quarter dwindled. Nonetheless, from mid-May until July 1941, the three Gladiators and a Cadet moved to a temporary strip at Bellinter House near Navan, Co Meath, observing suspicious troop activities in the north. When the North Strand was struck the Gladiators weren't called up even as a gesture.

In 1941, too, despite the waning possibility of German invasion, provision of emergency airfields for possible dispersed Air Corps or RAF use was again considered. Two leading sites emerged, at Gaybrook, Co Westmeath, and Rathduff, Golden, Co Tipperary. Rathduff was selected, work starting in autumn that year; by December two runways were almost complete.

A Lysander is refuelled from a mobile bowser, spat panels removed to prevent its mainwheels clogging with mud. In the foreground, the spat of another example.

Hectors 78 and 85 at Baldonnel in 1941 or 1942; a groundcrew member is putting his weight on 78's tail.

Dublin's North Strand area the morning after the *Luftwaffe* bombing of 30/31 May 1941.

Above: **Camouflaged Magister 35 at rest. Delivered to Baldonnel in March 1939, it served with the Training Schools.**

Left: **'Right men, gather round.' Using a Lysander as his teaching aid, an instructor imparts knowledge to attentive trainees.**

Magister 34 after a mishap, its undercarriage severely damaged and the propeller fractured. Repaired, the machine served until 1952; today it's exhibited at the National Museum of Ireland.

Above left: DH.84 Dragon II 18. On 16 December 1941, this machine crashed at Baldonnel and was wrecked. The pilot survived, while 18's target-towing apparatus was later used by windfall Battle 92.

Above right: An anonymous Hector tipped up after a landing accident. A recovery crew and Coles crane stand by.

Below: Walrus 20 at Fermoy, its hull national marking set within a white square. During the Blackwater manoeuvres, on 3 September 1942 this aircraft crashed at Castletownroche, Co Cork, following engine failure. The machine was written off but its pilot, Lt T P O'Mahony, lived to tell the tale.

Unplanned Arrivals

During the war years, more than 180 Allied and German aeroplanes alighted, force-landed or crashed in Ireland. Estimates vary but bad weather, navigational error, low fuel, mechanical problems and battle damage all contributed. Some machines were still flyable, while recovery of non-airworthy aircraft became an Air Corps responsibility, under the direction of G2 Branch (the army's intelligence service). If they were too damaged to warrant salvage or located in inaccessible areas, sometimes useful parts were recovered.

Hudson I P5123 was taken into service by Reconnaissance and Medium Bombing Squadron as 91. Behind is the RAF's 231 Squadron Tomahawk AH920, which had landed at Kilcarra, Co Wicklow, after becoming lost.

Battle TT.I V1222 under army guard after its arrival at Tramore, near Waterford.

Hurricane 93 flew with the Training Schools' Advanced Training Section and later with Fighter Squadron, serving until August 1945.

Windfall Hurricane 95 with Air Corps pupils, and an instructor in civvies, autumn 1942.

Of the RAF aircraft that arrived, several were repaired, purchased from Britain and taken into service. Hawker Hurricane I P5178 of 79 Squadron landed wheels-up on 29 September 1940 at Kilmuckridge, Ballyvadden, Co Wexford. On 21 December, 307 Squadron's Miles Master I N8009 crash-landed at Dungooley, Co Louth, following navigational error. Short of fuel, on 24 January 1941, Lockheed Hudson I P5123 of 233 Squadron alighted at Skreen, Co Sligo, while on 24 April, Fairey Battle TT.I V1222 belonging to West Freugh's 4 Bombing & Gunnery School came down at Tramore near Waterford during a training flight.

The Hudson joined the Reconnaissance and Medium Bombing Squadron, its new identity 91. As 92, the Battle became a target tug with the Training Schools, acquiring Donald Duck nose art and serving until May 1946. The Hurricane, its purchase marked by haggling (Mulcahy felt the £7,000 asking price was steep for a damaged machine), became 93. Initially 96, the Master was found to have a cracked wing spar so was used instructionally, becoming airframe A5.

On 10 June 1941, 32 Squadron's Hurricane IIA Z2832 landed with dry tanks at Whitestown, Co Waterford. During a delivery flight, on 21 August that year, Hurricane IIB Z5070 came down at Athboy, Co Meath. Following talks with Britain these aircraft also joined up, Z2832 becoming 94 and Z5070 95, both serving with the Training Schools' Advanced Training Section.

Heinkel He 111H-5 1G+LH of I./KG.27 crash-landed at Bunmahon, Co Waterford, on 1 April 1941; its five occupants were interned. Soldiers survey the wreck.

On 15 January 1943, while ferrying senior US Army officers from Gibraltar to Portreath, wandering Boeing B-17E 41-9045 *Stinky* force-landed at Athenry's Mellows Agricultural College, Co Galway. Crew and passengers were discreetly dispatched to Northern Ireland.

Lost during a flight from Morocco to Britain, on 5 November 1943, Douglas C-47A-40-DL Skytrain 42-24074 of 435th TCG, USAAF, alighted at Rineanna. American and Irish personnel pose for their photograph.

Skibbereen, Co Cork, April 1944: RAF (Met) Squadron 517. Handley Page Halifax LL145 is dismantled after its unintended arrival following an Atlantic weather reconnaissance flight. Rearing over salvage personnel, the machine has lost its nose, engines and wing leading edges. It was later taken by truck to Northern Ireland.

Internment

Built in late 1939 on Curragh Camp's east side, No.2 Internment Camp housed belligerent nations' military personnel who'd arrived on Irish soil and hadn't been able or allowed to return home. These men consisted of flyers, and seafarers from Germany's *Kriegsmarine*. The site was also known as K-Lines, its name derived from one of Curragh Camp's original alphabetically-arranged square areas. Allied and German internees, of course, lived in separate areas. Hostility was generally limited, but occasionally the two groups met while visiting the local area on parole.

On 25 November 1940, *Utffz* Wilhelm Krupp's Blohm und Voss Bv138A-1 of 2/Kflgp906 had alighted with engine problems near Innisvickillane island, off Co Kerry. Spitfire Vb BM533 and Polish Sgt Jan Zimek of 41 Squadron had force-landed at Wells, Co Wexford, on 31 October 1942. Both men entered K-Lines. At a dance at nearby Newbridge town in May 1943, a clash occurred when Zimek approached Krupp, calling him 'a fucking Hun'. Krupp swung out, but a local man restrained the pair; later that night Krupp had another go. In July, Zimek was repatriated to the north.

In fact, rather than internment, during the Emergency more than 250 downed Allied airmen were returned to the north. By June 1944, all those detained had been released. German flyers though, eventually over 50 men, together with the *Kriegsmarine* personnel, remained until the war's end. During his time at K-Lines, *Utffz* Ulrich Winkler, ex-crewman of Focke-Wulf Fw 200 C-6 Condor F8+MR, which had come down near Nenagh, Tipperary, on 13 December 1943, mended Newbridge church's organ. Some 40 years later, on holiday in the area he called there; his repairs were still going strong.

Curragh Camp's No.2 Internment Camp, also known as K-Lines; *Luftwaffe* and *Kriegsmarine* personnel line up for the camera.

A Word on Neutrality

Despite Ireland's neutrality, during the Emergency veiled assistance to Britain grew in numerous ways, including aviation-related matters. Information on Rineanna's weather observations and German sightings was made available to the British. By early 1941, following the Dáil's covert agreement to the creation of a flight path known as the Donegal Corridor, RAF flying-boats from Lough Erne in the north were permitted to traverse a narrow four-mile strip of Irish territory between Belleek, Co Fermanagh, and Ballyshannon, Co Donegal. This allowed shorter access to the Atlantic and thereby greater reach for maritime patrols.

Over the Emergency too, around 39 British and American aircraft landed intact in Ireland, and were refuelled and allowed to return home, often benevolently interpreted as lost during peaceful missions: air-sea rescues, ferry flights, air tests. Nearly 30 machines, damaged while landing, were quietly transported across the border.

Certainly, Emergency-period aircraft purchases from Britain were influenced by overarching diplomatic talks between the two nations on the extent of Ireland's shrouded support despite her non-alignment. Against that backdrop, Col Mulcahy continued to press for more machines. He cultivated Britain's air attaché to Dublin, incognito Wg Cdr Ralph Lywood, showing him round Baldonnel, taking him up in an Anson to observe various airfields and possible sites for the emergency airstrips, and passing intelligence including details of the strips.

For some time, of course, Britain's own resources were desperately hard-pressed. But even when fortunes improved, supply of aeroplanes (particularly fighting types) remained a sluggish trickle. This was due not least to lingering uncertainty over Ireland's international sympathies, regardless of her neutral status, and despite her flexibility toward the Allies under – to say the least – difficult circumstances.

During the Emergency, private flying in Ireland was suspended. In May 1940, Robert Albert Clarke's Spartan Three Seater II EI-ABU was grounded at Cloughjordan sawmill, Co Tipperary, its rudder and elevators removed. However, it survived and today is New Zealand-based as ZK-ARH. Plans are currently afoot to return the machine to the Isle of Wight, where it was built in 1932.

Left: Co Wexford's Hook Lighthouse with its stone Emergency-era sign, 16 EIRE, 1944. To help indicate neutral territory, more than 80 such markers appeared, spread round much of Ireland's coastline.

Below: Delivered in April 1940 from Fokker, the first Aer Lingus DC-3, -268B variant EI-ACA, at Dublin airport (formerly the Collinstown site) wearing Emergency camouflage. During the war years, Aer Lingus services were mainly internal, but with some flights to Liverpool or Manchester depending on German activity. On 18 June 1946, EI-ACA was written off, without fatalities, when a port engine fire shortly after take-off led to a forced landing near Shannon airport.

Finally, More Aircraft

During 1941, a top-level enquiry was convened by army Chief of Staff Lt Gen Daniel McKenna, Mulcahy's senior, to examine the effectiveness, organisation, equipment, training and administration of the Air Corps. It reported in January 1942, unsurprisingly calling for more pilots, machines and flying training, as well as two new reconnaissance and five fighter squadrons. However, although some additional pilots appeared, interceptors remained few. This was underscored when Hinds 67, 68 and 72 moved to Fighter Squadron that year, serving for around 12 months. The seven requested squadrons never materialised.

Right: During the Blackwater manoeuvres, on 5 September 1942 Magister 35 from Glenville crashed and was written off, though pilot Lt D V Cousins survived. Amid the harvest, two ladies and a military gent pose with the wreck.

Below: Hector 88 came to grief at Rathduff on 3 September 1942, being withdrawn from use after just nine months' service.

On 9 January 1942, Walrus 18 made a unique journey with Lt Alphonsus 'Alan' Thornton and three confederates. Restless at Rineanna, with limited flying and only The Honk public house for off-duty amusement, the men had decided life would be more stimulating with the *Luftwaffe*. Heading for France to volunteer, the Walrus evaded a searching Lysander but toward the British coastline was intercepted by four RAF Spitfires, and compelled to land at Cornwall's St Eval airfield.

The Irish authorities were contacted, the men detained and subsequently taken home under guard. Collected by two Air Corps pilots, 18 returned to Baldonnel sporting a gleaming new Lewis gun donated by St Eval's armourers, though the War Office charged 10s 10d to supply fuel and oil for the flight. Thornton served a prison sentence before changing his plans, joining the RAF and serving in the Mediterranean theatre.

September 1942: three Hectors and a Lysander at Rathduff during the Blackwater manoeuvres. Hector 78 had served since May 1941, being withdrawn in October 1943.

On parade: Hurricane Is at Baldonnel. Left and centre examples wear Fighter Squadron badges; centre is 106.

In August and September, Air Corps machines participated in the Blackwater manoeuvres, substantial army exercises. Though there was no longer any real prospect of invasion, these sought to assess the armed forces' capacity to oppose a seaborne landing in the south-east. Aircraft from Rineanna and Rathduff supported, respectively, the army's 1st and 2nd Divisions. Fermoy and a strip at Glenville, Co Cork, were also used, flying performed wholly during daylight though in varied weather. Types involved included Ansons, Cadets, Hectors, Lysanders and Magisters, reporting ground movements as best they could (wirelesses weren't universal) while in turn the army endeavoured to avoid observation.

⊹ ⊹ ⊹

In February 1943, six ex-RAF stock Miles Master IIs arrived at Baldonnel, in part fruits of Mulcahy's overtures to Wg Cdr Lywood, assuming 97–102. Never having entered service, they were in good shape. The following month, with an Air Corps reorganisation, Maj William Delamere (who'd served since September 1922) became its commanding officer. During April, Fighter Squadron moved to Rineanna, where facilities had slowly improved. Coastal patrols used Hurricane 93, the three Gladiators and remaining Lysanders 61, 63 and 66. Masters 97, 98 and 99 also travelled west, the others joining the Training Schools.

As for additional fighting aircraft, in July, Fighter Squadron at last received four ex-RAF Hurricane Is (allotted 103–106), though all were markedly second-hand. That month too, Hurricanes 94 and 95 were returned to Britain. During November, three more Hurricane Is (107–109) arrived; a single example (110) appeared in February 1944 and three (111, 112 and 114) in March. Curiously, it seems 110 was never formally taken on to Fighter Squadron's books.

September 1944 saw Lysanders 61 and 66 converted to target tugs by Short and Harland in Belfast. Their stub bomb-carriers removed, repainted silver overall and redesignated TT.IIs, they served until 1946. In reduced circumstances 66 survived until the 1960s, kept outside Crookstown Motors at Athy, Co Kildare, until its remains were scrapped.

Master II 98 at Baldonnel with the five other Air Corps examples, soon after delivery in February 1943. Its previous owner's serial, W9028, is still present on the fuselage and underwing.

During March 1945 six Hurricane IICs were delivered, assuming 115–120, the last aeroplanes acquired before the war in Europe ended. In May, Fighter Squadron transferred to Gormanston. On 5 May, a fugitive from Grove airfield, Denmark, also arrived there: *Luftwaffe* Junkers Ju 88 G-6 D5+GH of I./NJG3. The Air Corps had no use for it and swapped it with Britain for aircraft spares and ammunition. Leading British test-pilot T/Lt (A) (later Capt) Eric Brown flew it from Gormanston to Farnborough for evaluation.

In May, the most terrible conflict in history finally came to an end. Air Corps flying duly reduced. Over that summer and autumn, the Royal Dublin Society held a military tattoo and exhibition at Ballsbridge, two Hurricanes and a Magister being displayed. In September the following year, the Emergency Powers Act lapsed. The aging Hurricanes were plagued by maintenance problems, but the final example (120) lingered until November 1947.

+ + +

Today there are still a few reminders of the Air Corps' earlier times. Cadet C7 is preserved at Baldonnel's Irish Air Corps Museum, and Magister 34 at the National Museum of Ireland's Collins Barracks. Parts of Hector 88 and another have survived, and at Baldonnel a Napier Dagger III. Walrus 18 was struck off in August 1945; following brief Aer Lingus service and subsequent abandonment, in 1963 it was acquired by the Historic Aircraft Preservation Society. Wearing its original identity L2301, it's now displayed at Yeovilton's Fleet Air Arm Museum. And in a revered place in the Baldonnel officers' mess, mounted in a glass case is *The Big Fella's* propeller hub.

Magister formation: left to right 75, 34, 74, 37, 32 and 39. On 15 May 1944, 37 crashed at Scribblestown, Co Dublin; pupil Sgt-Pilot Arthur Casserly was killed.

Master II 101 and to its right 97, by the civil terminal at Collinstown (now Dublin Airport). On 19 August 1946, 101 crashed at Castlebaggott farmhouse near Baldonnel, killing pilot Capt Maurice Quinlan.

Right: Hurricane Is at Baldonnel, 30 March 1944: 110, 105, 103, 111, 112 and 114 are identifiable, 105 displaying Fighter Squadron's badge, while 110 sports the type's longer Rotol spinner.

Below: Cannon-armed Hurricane IICs 116, 115 and 119 out for a photoshoot. The longest lived of these was 116, which served until July 1947.

Left: May 1945, Gormanston: ex-I./NJG3 Junkers Ju 88 G-6 D5+GH with three of its aircrew.

Below: The Royal Dublin Society's 1945 exhibition included Magister 34, Hurricane I 104 and Hurricane IIC 120, together with a Link trainer.

The Emergency, 1939–1945

Above: Wings Day, 22 June 1946, the 4th Young Officers Wings Class: Left to right, 2/Lts W J Glenn, E F Farrelly, P O Ferguson, J Liddy, G McCabe, H Houston, J L M Connolly, F Young. Four Ansons survived the wartime years: in the background is (probably) 42, its turret unarmed.

Below left: Preserved Miles Magister 34 during its time at Baldonnel's Irish Air Corps Museum.

Below right: 'Small but Fierce': today No.1 Fighter Squadron's badge and motto, *Beag ach Fiachmhar*, are depicted in a stained-glass window at Baldonnel's church, Our Lady of Loreto and St Brigid. The badge was designed by Lts Des Johnston and Andrew Woods.

Appendix 1
Ex-RAF identities of Air Service aircraft, Irish Civil War

Bristol F.2B: B.I, ex-E2411; B.II, ex-H1251; B.III, ex-H1485; IV, ex-E1958; V, ex-D7865; VI, ex-D7886; BF.VII, ex-D7882; VIII, ex-D7885.

Avro 504K: I, ex-H2500; II, ex-H2073; III, ex-H2075; IV, ex-E359; V, ex-H2505; VI, ex-D7588 (and G-EADQ).

RAF S.E.5a: II, ex-F5282.

Martinsyde F.4: I, ex-D4285; II, ex-D4281; III, ex-D4298; IV, ex-D4274.

Airco DH.9: I, ex-H5797; II, ex-H5830; III, ex-H5774; IV, ex-H5869; V, ex-H5823; VI, ex-H9310; 7, ex-H9247; 8, ex-H5862.

Appendix 2

Ex-RAF identities of Air Corps aircraft, Emergency period corresponding with the World War Two in Europe

Avro Anson I: 41, ex-N4863; 42, ex-N4864; 43, ex-N4865; 44, ex-N4866; 45, ex-N4867.

Miles Magister I (first batch): 31, ex-N5389; 32, ex-N5390; 33, ex-N5391; 34, ex-N5392; 35, ex-N5393; 36, ex-N5400; 37, ex-N5401; 38, ex-N5402; 39, ex-N5403; 40, ex-N5404. Diverted from an RAF order, none had entered service.

Supermarine Walrus I: N18, ex-L2301; N19, ex-L2302; N20, ex-L2303. Diverted from Fleet Air Arm order, none had entered service.

Hawker Hind I: 67, ex-K5446; 68, ex-K5559; 69, ex-K6712; 70, ex-K5415; 71, ex-K6755; 72, ex-K6781.

Miles Magister I (second batch): 73, ex-L6903; 74, ex-P6440; 75, ex-N3901; 76, ex-P6414; 77, ex-P6422.

Hawker Hector I: 78, ex-K8098; 79, ex-K8102; 80, ex-K8105; 81, ex-K8114; 82, ex-K8115; 83, ex-K8117; 84, ex-K8148; 85, ex-K9697; 86, ex-K9725; 87, ex-K9715; 88, ex-K8130; 89, ex-K8159; 90, ex-K9761.

Lockheed Hudson I: 91, ex-P5123.

Fairey Battle TT.1: 92, ex-V1222.

Hawker Hurricane I: 93, ex-P5178.

Miles Master I: 96, ex-N8009 (to instructional airframe A5).

Hawker Hurricane IIA: 94, ex-Z2832.

Hawker Hurricane IIB: 95, ex-Z5070.

Miles Master II: 97, ex-DM260; 98, ex-W9028; 99, ex-DM258; 100, ex-DL352; 101, ex-AZ741; 102, ex-DM261. Diverted from RAF orders; none had entered service.

Hawker Hurricane I: 103, ex-V6613; 104, ex-V7411; 105, ex-V7540; 106, ex-Z4037; 107, ex-P2968; 108, ex-P3416; 109, ex-V7173; 110, ex-Z7158; 111, ex-V6576; 112, ex-V7435; 114, ex-V7463.

Hawker Hurricane IIC: 115, ex-LF536; 116, ex-LF541; 117, ex-LF566; 118, ex-LF624; 119, ex-LF770; 120, ex-PZ796.

Other books you might like:

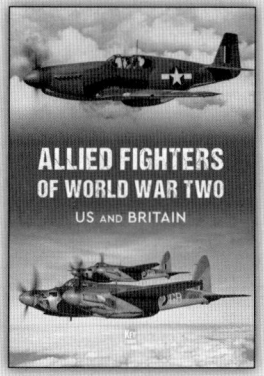

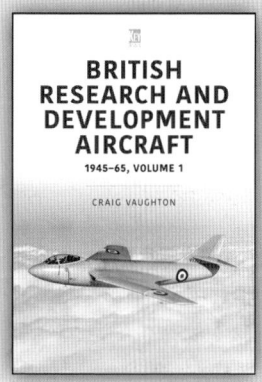
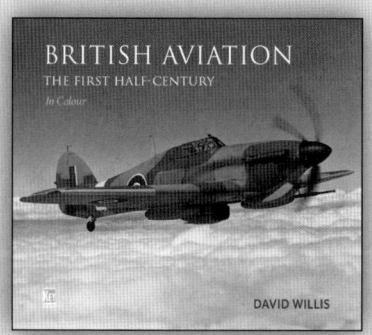
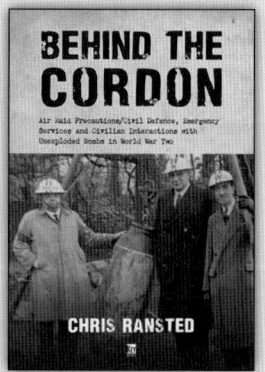

For our full range of titles please visit:
shop.keypublishing.com/books

VIP Book Club

Sign up today and receive
TWO FREE E-BOOKS

Be the first to find out about our forthcoming book releases and receive exclusive offers.

Register now at **keypublishing.com/vip-book-club**

Our VIP Book Club is a 100% spam-free zone, and we will never share your email with anyone else. You can read our full privacy policy at: privacy.keypublishing.com